Advanced Applied Mathematics

Advanced Applied Mathematics

Rohit Gupta

Oxford Book Company
Jaipur ▪ New Delhi

Published By

Oxford Book Company

Regd. Off. "Bony Residency", Gate No. 2
Opp. Tilak Public School, Vishweswariya Nagar,
Gopalpura Road, Jaipur - 302018, India
Ph: 0141 - 2761280, Telefax: 2761381
email - oxfordbookjpr@gmail.com
website - oxfordbookcompany.com

Delhi Office: 102, 1st Floor, Satyam House,
4327/3, Ansari Road, Darya Ganj,
New Delhi - 110002, India
Ph: 011-45652440

Distributed By
KEYA INTERNATIONAL PRESS LLP
E-mail: keyainternationalpressllp@gmail.com

First published in India, 2026

ISBN : 978-93-5524-529-8

Typeset by Shivangi Computers, Jaipur

Printed and bound in India by Thomson Press Ltd.,
New Delhi

Contents

Preface

Mathematics is the universal language of science and technology, providing the essential framework for solving complex problems in diverse fields. Applied mathematics, in particular, serves as a bridge between abstract mathematical theory and real-world applications. This book, "Advanced Applied Mathematics," is designed to equip students, researchers, and professionals with the analytical tools and mathematical techniques required to tackle advanced problems in engineering, physics, economics, and computational sciences.

The study of applied mathematics extends beyond routine calculations—it involves modeling real-world phenomena, analyzing intricate systems, and developing efficient algorithms. This book delves into advanced topics such as differential equations, numerical analysis, optimization, functional analysis, and mathematical modeling. Each concept is introduced with its theoretical foundation, followed by practical applications that demonstrate its relevance in modern scientific and industrial domains.

A distinctive feature of this book is its problem-solving approach, which balances mathematical rigor with practical insights. The theoretical discussions are complemented by numerous examples, case studies, and problem sets designed to test comprehension and foster deeper engagement with the subject matter. Whether it is optimizing resource allocation, analyzing stability in dynamical systems, or solving partial differential equations, the reader will find a structured methodology to approach complex mathematical challenges.

The material presented in this book is suitable for advanced undergraduate and graduate students, as well as professionals seeking to enhance their mathematical proficiency. Unlike conventional texts that focus solely on theory, this book emphasizes application-driven learning, ensuring that readers not only understand mathematical principles but also develop the skills to implement them in real-world scenarios.

Moreover, special attention has been given to modern computational techniques that are widely used in scientific computing and engineering analysis. With the advent of high-performance computing, applied mathematics has evolved significantly, necessitating a blend of traditional mathematical techniques and computational methods. This book integrates both perspectives, offering a well-rounded approach to contemporary applied mathematics.

— ***Rohit Gupta***

Chapter 1

Integral Methods of Calculus

Integration is an important concept in mathematics and, together with its inverse, differentiation, is one of the two main operations in calculus. Given a function f of a real variable x and an interval $[a, b]$ of the real line, the definite integral

$$\int_a^b f(x)dx$$

is defined informally to be the area of the region in the xy-plane bounded by the graph of f, the x-axis, and the vertical lines $x = a$ and $x = b$, such that areas above the axis add to the total, and the area below the x axis subtract from the total.

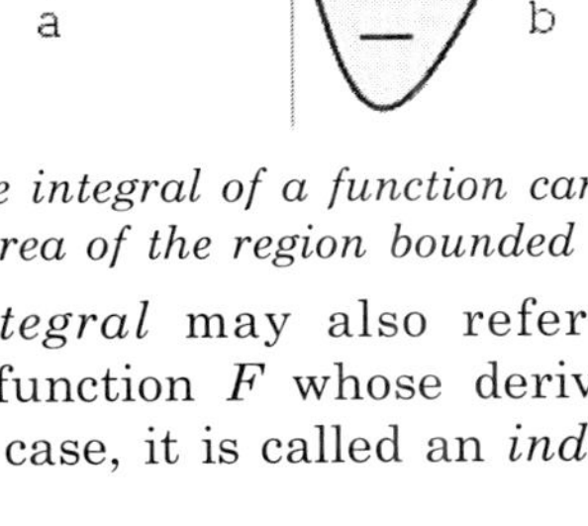

Figure: *A definite integral of a function can be represented as the signed area of the region bounded by its graph.*

The term *integral* may also refer to the notion of antiderivative, a function F whose derivative is the given function f. In this case, it is called an *indefinite integral* and is written:

$$F = \int f(x)dx.$$

The integrals discussed in this article are termed *definite integrals*. The principles of integration were formulated independently by Isaac Newton and Gottfried Leibniz in the late 17th century. Through the fundamental theorem of calculus, which they independently developed, integration is connected with differentiation: if f is a continuous real-valued function defined on a closed interval $[a, b]$, then, once an antiderivative F of f is known, the definite integral of f over that interval is given by

$$\int_a^b f(x)dx = F(b) - F(a)$$

Integrals and derivatives became the basic tools of calculus, with numerous applications in science and engineering. The founders of the calculus thought of the integral as an infinite sum of rectangles of infinitesimal width. A rigorous mathematical definition of the integral was given by Bernhard Riemann. It is based on a limiting procedure which approximates the area of a curvilinear region by breaking the region into thin vertical slabs. Beginning in the nineteenth century, more sophisticated notions of integrals began to appear, where the type of the function as well as the domain over which the integration is performed has been generalised. A line integral is defined for functions of two or three variables, and the interval of integration $[a, b]$ is replaced by a certain curve connecting two points on the plane or in the space. In a surface integral, the curve is replaced by a piece of a surface in the three-dimensional space. Integrals of differential forms play a fundamental role in modern differential geometry. These generalizations of integrals first arose from the needs of physics, and they play an important role in the formulation of many physical laws, notably those of electrodynamics. There are many modern concepts of integration, among these, the most common is based on the abstract mathematical theory known as Lebesgue integration, developed by Henri Lebesgue.

Pre-calculus Integration

The first documented systematic technique capable of determining integrals is the method of exhaustion of the ancient Greek astronomer Eudoxus (*ca.* 370 BC), which sought to find areas and volumes by breaking them up into an infinite number of shapes for which the area or volume was known. This method was further developed and employed by Archimedes in the 3rd century BC and used to calculate areas for parabolas and an approximation to the area of a circle. Similar methods were independently developed in China around the 3rd century AD by Liu Hui, who used it to find the area of the circle. This method was later used in the 5th century by Chinese father-and-son mathematicians Zu Chongzhi and Zu Geng to find the volume of a sphere (Shea 2007; Katz 2004, pp. 125–126).

The next significant advances in integral calculus did not begin to appear until the 16th century. At this time the work of Cavalieri with his *method of indivisibles*, and work by Fermat, began to lay the foundations of modern calculus, with Cavalieri computing the integrals of x^n up to degree $n = 9$ in Cavalieri's quadrature formula. Further steps were made in the early 17th century by Barrow and Torricelli, who provided the first hints of a connection between integration and differentiation. Barrow provided the first proof of the fundamental theorem of calculus. Wallis generalized Cavalieri's method, computing integrals of x to a general power, including negative powers and fractional powers.

Newton and Leibniz

The major advance in integration came in the 17th century with the independent discovery of the fundamental theorem of calculus by Newton and Leibniz. The theorem demonstrates a connection between integration and differentiation. This connection, combined with the comparative ease of differentiation, can be exploited to calculate integrals. In particular, the fundamental theorem of calculus allows one to solve a much broader class of problems. Equal in importance is the comprehensive mathematical framework that both Newton and Leibniz developed. Given the name infinitesimal calculus, it allowed for precise analysis of functions within continuous domains. This framework eventually became modern calculus, whose notation for integrals is drawn directly from the work of Leibniz.

Formalizing Integrals

While Newton and Leibniz provided a systematic approach to integration, their work lacked a degree of rigour. Bishop Berkeley memorably attacked the vanishing increments used by Newton, calling them "ghosts of departed quantities". Calculus acquired a firmer footing with the development of limits. Integration was first rigorously formalized, using limits, by Riemann. Although all bounded piecewise continuous functions are Riemann integrable on a bounded interval, subsequently more general functions were considered – particularly in the context of Fourier analysis – to which Riemann's definition does not apply, and Lebesgue formulated a different definition of integral, founded in measure theory (a subfield of real analysis). Other definitions of integral, extending Riemann's and Lebesgue's approaches, were proposed. These approaches based on the real number system are the ones most common today, but alternative approaches exist, such as a definition of integral as the standard part of an infinite Riemann sum, based on the hyperreal number system.

Historical Notation

Isaac Newton used a small vertical bar above a variable to indicate integration, or placed the variable inside a box. The vertical bar was easily confused with $\dot{x}$ or x', which Newton used to indicate differen-tiation, and the box notation was difficult for printers to reproduce, so these notations were not widely adopted. The modern notation for the indefinite integral was introduced by Gottfried Leibniz in 1675 (Burton 1988, p. 359; Leibniz 1899, p. 154). He adapted the integral symbol, *ſ*, from the letter *ſ* (long s), standing for *summa* (written as *numma*; Latin for "sum" or "total"). The modern notation for the definite integral, with limits above and below the integral sign, was first used by Joseph Fourier in *Mémoires* of the French Academy around 1819–20, reprinted in his book of 1822.

Terminology and Notation

The simplest case, the integral over x of a real-valued function $f(x)$, is written as

$$\int f(x)dx.$$

The integral sign *ſ* represents integration. The dx indicates that we are integrating over x; dx is called the variable of integration. In correct mathematical typography, the dx is separated from the integrand by a space (as shown). Some authors use an upright d (that is, dx instead of dx). Inside the *ſ*...dx is the expression to be integrated, called the *integrand*. In this case the integrand is the function $f(x)$. Because there is no domain specified, the integral is called an *indefinite integral*.

When integrating over a specified domain, we speak of a *definite integral*. Integrating over a domain D is written as

$$\int_D f(x)dx, \text{or } \int_a^b f(x)dx$$

if the domain is an interval $[a, b]$ of x.

The domain D or the interval $[a, b]$ is called the *domain of integration*. If a function has an integral, it is said to be *integrable*. In general, the integrand may be a function of more than one variable, and the domain of integration may be an area, volume, a higher dimensional region, or even an abstract space that does not have a geometric structure in any usual sense (such as a sample space in probability theory). In the modern Arabic mathematical notation, which aims at pre-university levels of education in the Arab world and is written from right to left, a reflected integral symbol ∫ is used (W3C 2006).

The variable of integration dx has different interpretations depending on the theory being used. It can be seen as strictly a notation indicating that x is a dummy variable of integration; if the integral is seen as a Riemann sum, dx is a reflection of the weights or widths d of the intervals of x; in Lebesgue integration and its extensions, dx is a measure; in non-standard analysis, it is an infinitesimal; or it can be seen as an independent mathematical quantity, a differential form. More complicated cases may vary the notation slightly.

In Leibniz's notation, dx is interpreted an infinitesimal change in x, but his interpretation lacks rigour in the end. Nonetheless Leibniz's notation is the most common one today; and as few people are in need of full rigour, even his interpretation is still used in many settings.

Integrals appear in many practical situations. If a swimming pool is rectangular with a flat bottom, then from its length, width, and depth we can easily determine the volume of water it can contain (to fill it), the area of its surface (to cover it), and the length of its edge (to rope it). But if it is oval with a rounded bottom, all of these quantities call for integrals.

Practical approximations may suffice for such trivial examples, but precision engineering (of any discipline) requires exact and rigorous values for these elements.

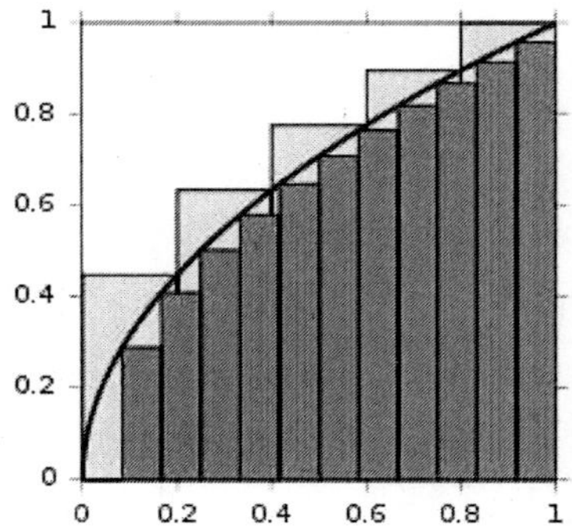

Approximations to integral of √*x* from 0 to 1, with % 5 right samples (above) and % 12 left samples (below). To start off, consider the curve $y = f(x)$ between $x = 0$ and $x = 1$ with $f(x) = \sqrt{x}$. We ask:

What is the area under the function *f*, in the interval from 0 to 1? and call this (yet unknown) area the integral of *f*. The notation for this integral will be

$$\int_0^1 \sqrt{x}\,dx.$$

As a first approximation, look at the unit square given by the sides $x = 0$ to $x = 1$ and $y = f(0) = 0$ and $y = f(1) = 1$. Its area is exactly 1. As it is, the true value of the integral must be somewhat less. Decreasing the width of the approximation rectangles shall give a better result; so cross the interval in five steps, using the approximation points 0, 1/5, 2/5, and so on to 1. Fit a box for each step using the right end height of each curve piece, thus √(1D†5), √(2D†5), and so on to √1 = 1. Summing the areas of these rectangles, we get a better approximation for the sought integral, namely

$$\sqrt{\frac{1}{5}}\left(\frac{1}{5}-0\right)+\sqrt{\frac{2}{5}}\left(\frac{2}{5}-\frac{1}{5}\right)+\cdots+\sqrt{\frac{5}{5}}\left(\frac{5}{5}-\frac{4}{5}\right)\approx 0.7497.$$

Notice that we are taking a sum of finitely many function values of *f*, multiplied with the differences of two subsequent approximation points. We can easily see that the approximation is still too large. Using more steps produces a closer approximation, but will never be exact: replacing the 5 subintervals by twelve as depicted, we will get an approximate value for the area of 0.6203, which is too small. The key idea is the transition from adding *finitely many* differences of approximation points multiplied by their respective function values to using infinitely many fine, or *infinitesimal* steps.

As for the *actual calculation of integrals*, the fundamental theorem of calculus, due to Newton and Leibniz, is the fundamental link between the operations of differentiating and integrating. Applied to the square root curve, $f(x) = x^{1/2}$, it says to look at the antiderivative $F(x) = (2/3)x^{3/2}$, and simply take $F(1) - F(0)$, where 0 and 1 are the boundaries of the interval [0,1]. So the *exact* value of the area under the curve is computed formally as

$$\int_0^1 \sqrt{x}\, dx = \int_0^1 x^{1/2}\, dx = F(1) - F(0) = 2/3.$$

(This is a case of a general rule, that for $f(x) = x^q$, with $q \neq -1$, the related function, the so-called antiderivative is $F(x) = x^{q+1}/(q + 1)$.)

The notation

$$\int f(x)dx$$

conceives the integral as a weighted sum, denoted by the elongated *s*, of function values, $f(x)$, multiplied by infinitesimal step widths, the so-called *differentials*, denoted by dx. The multiplication sign is usually omitted. Historically, after the failure of early efforts to rigorously interpret infinitesimals, Riemann formally defined integrals as a limit of weighted sums, so that the dx suggested the limit of a difference (namely, the interval width). Shortcomings of Riemann's dependence on intervals and continuity motivated newer definitions, especially the Lebesgue integral, which is founded on an ability to extend the idea of "measure" in much more flexible ways. Thus the notation

$$\int_A f(x)d\mu$$

refers to a weighted sum in which the function values are partitioned, with μ measuring the weight to be assigned to each value. Here A denotes the region of integration. Differential geometry, with its "calculus on manifolds", gives the familiar notation yet another interpretation. Now $f(x)$ and dx become a differential form, $\omega = f(x)\ dx$, a new differential operator d, known as the exterior derivative is introduced, and the fundamental theorem becomes the more general Stokes' theorem,

$$\int_A d\omega = \int_{\partial A} \omega,$$

from which Green's theorem, the divergence theorem, and the fundamental theorem of calculus follow. More recently, infinitesimals have reappeared with rigor, through modern innovations such as non-standard analysis. Not only do these methods vindicate the intuitions of the pioneers; they also lead to new mathematics. Although there are differences between these conceptions of integral, there is considerable overlap. Thus, the area of the surface of the oval swimming pool can be handled as a geometric ellipse, a sum of infinitesimals, a Riemann integral, a Lebesgue integral, or as a manifold with a differential form. The calculated result will be the same for all.

Formal Definitions

There are many ways of formally defining an integral, not all of which are equivalent. The differences exist mostly to deal with differing special cases which may not be integrable under other definitions, but also occasionally for pedagogical reasons. The most commonly used definitions of integral are Riemann integrals and Lebesgue integrals.

Other Integrals

Riemann–Stieltjes Integral

In mathematics, the Riemann–Stieltjes integral is a generalization of the Riemann integral, named after Bernhard Riemann and Thomas Joannes Stieltjes.

Definition

The Riemann–Stieltjes integral of a real-valued function f of a real variable with respect to a real function g is denoted by

$$\int_a^b f(x)dg(x)$$

and defined to be the limit, as the mesh of the partition

$$P = \{a = x_0 < x_1 < \cdots < x_n = b\}$$

of the interval $[a, b]$ approaches zero, of the approximating sum

$$S(P, f, g) = \sum_{i=0}^{n-1} f(c_i)(g(x_{i+1}) - g(x_i))$$

where c_i is in the i-th subinterval $[x_i, x_{i+1}]$. The two functions f and g are respectively called the integrand and the integrator.

The "limit" is here understood to be a number A (the value of the Riemann–Stieltjes integral) such that for every $\varepsilon > 0$, there exists $\delta > 0$ such that for every partition P with mesh$(P) < \delta$, and for every choice of points c_i in $[x_i, x_{i+1}]$,

$$|S(P, f, g) - A| < \varepsilon.$$

Generalized Riemann–Stieltjes Integral

A slight generalization, introduced by Pollard (1920) and now standard in analysis, is to consider in the above definition partitions P that *refine* another partition P_ε, meaning that P arises from P_ε, by the addition of points, rather than from partitions with a finer mesh. Specifically, the generalized Riemann–Stieltjes integral of f with respect to g is a number A such that for every $\varepsilon > 0$ there exists a partition P_ε such that for every partition P that refines P_ε,

$$|S(P, f, g) - A| < \varepsilon$$

for every choice of points c_i in $[x_i, x_{i+1}]$.

This generalization exhibits the Riemann–Stieltjes integral as the Moore–Smith limit on the directed set of partitions of $[a, b]$ (McShane 1952). Hildebrandt (1938) calls it the Pollard–Moore–Stieltjes integral.

Darboux Sums

The Riemann–Stieltjes integral can be efficiently handled using an appropriate generalization of Darboux sums. For a partition P and a nondecreasing function g on $[a, b]$ define the upper Darboux sum of f with respect to g by

$$U(P,f,g)=\sum_{i=1}^{n}\sup_{x\in[x_i,x_{i+1}]} f(x)(g(x_{i+1})-g(x_i))$$

and the lower sum by

$$L(P,f,g)=\sum_{i=1}^{n}\inf_{x\in[x_i,x_{i+1}]} f(x)(g(x_{i+1})-g(x_i)).$$

Then the generalized Riemann–Stieltjes of f with respect to g exists if and only if, for every $\varepsilon > 0$, there exists a partition P such that

$$U(P,f,g)-L(P,f,g)<\varepsilon.$$

Furthermore, f is Riemann–Stieltjes integrable with respect to g (in the classical sense) if

$$\lim_{\text{mesh}(P)\to 0}[U(P,f,g)-L(P,f,g)]=0.$$

Properties and Relation to the Riemann Integral

If g should happen to be everywhere differentiable, then the Riemann–Stieltjes integral may still be different from the Riemann integral of $f(x)g'(x)$ given by

$$\int_a^b f(x)g'(x)dx,$$

for example, if the derivative is unbounded. But if the derivative is continuous, they will be the same. This condition is also satisfied if g is the (Lebesgue) integral of its derivative; in this case g is said to be absolutely continuous.

However, g may have jump discontinuities, or may have derivative zero *almost* everywhere while still being continuous and increasing (for example, g could be the Cantor function), in either of which cases the Riemann–Stieltjes integral is not captured by any expression involving derivatives of g.

The Riemann–Stieltjes integral admits integration by parts in the form

$$\int_a^b f(x)dg(x)=f(b)g(b)-f(a)g(a)-\int_a^b g(x)df(x).$$

and the existence of either integral implies the existence of the other (Hille & Phillips 1974, §3.3).

Existence of the Integral

The best simple existence theorem states that if f is continuous and g is of bounded variation on $[a, b]$, then the integral exists. A function g is of bounded variation if and only if it is the difference between two monotone functions. If g is not of bounded variation, then there will be continuous functions which cannot be integrated with respect to g. In general, the integral is not well-defined if f and g share any points of discontinuity, but this sufficient condition is not necessary.

On the other hand, a classical result of Young (1936) states that the integral is well-defined if f is α–Hölder continuous and g is β-Hölder continuous with $\alpha + \beta > 1$.

Application to Probability Theory

If g is the cumulative probability distribution function of a random variable X that has a probability density function with respect to Lebesgue measure, and f is any function for which the expected value $E(|f(X)|)$ is finite, then the probability density function of X is the derivative of g and we have

$$E(f(X)) = \int_{-\infty}^{\infty} f(x)g'(x)dx.$$

But this formula does not work if X does not have a probability density function with respect to Lebesgue measure. In particular, it does not work if the distribution of X is discrete (i.e., all of the probability is accounted for by point-masses), and even if the cumulative distribution function g is continuous, it does not work if g fails to be absolutely continuous (again, the Cantor function may serve as an example of this failure). But the identity

$$E(f(X)) = \int_{-\infty}^{\infty} f(x)dg(x)$$

holds if g is *any* cumulative probability distribution function on the real line, no matter how ill-behaved. In particular, no matter how ill-behaved the cumulative distribution function g of a random variable X, if the moment $E(X^n)$ exists, then it is equal to

$$E(X^n) = \int_{-\infty}^{\infty} x^n dg(x).$$

Application to Functional Analysis

The Riemann–Stieltjes integral appears in the original formulation of F. Riesz's theorem which represents the dual space of the Banach space $C[a,b]$ of continuous functions in an interval $[a,b]$ as Riemann–Stieltjes integrals against functions of bounded variation. Later, that theorem was reformulated in terms of measures. The Riemann–Stieltjes integral also appears in the formulation of the spectral theorem for (non-compact) self-adjoint (or more generally, normal) operators in a Hilbert space. In this theorem, the integral is considered with respect to a spectral family of projections.

Generalization

An important generalization is the Lebesgue–Stieltjes integral which generalizes the Riemann–Stieltjes integral in a way analogous to how the Lebesgue integral generalizes the Riemann integral. If improper Riemann–Stieltjes integrals are allowed, the Lebesgue integral is not strictly more general than the Riemann–Stieltjes integral. The Riemann–Stieltjes integral also generalizes to the case when either the integrand f or the integrator g take values in a Banach space. If $g : [a,b] \to X$ takes values in the Banach space X, then it is natural to assume that it is of strongly bounded variation, meaning that

$$\sup \sum_i \| g(t_i) - g(t_{i+1}) \|_X < \infty$$

the supremum being taken over all finite partitions

$$a = t_0 \le t_1 \le \cdots \le t_n = b$$

of the interval $[a,b]$. This generalization plays a role in the study of semigroups, via the Laplace–Stieltjes transform.

Lebesgue–Stieltjes Integration

In measure-theoretic analysis and related branches of mathematics, Lebesgue–Stieltjes integration generalizes Riemann–Stieltjes and Lebesgue integration, preserving the many advantages of the former in a more general measure-theoretic framework. The Lebesgue–Stieltjes integral is the ordinary Lebesgue integral with respect to a measure known as the Lebesgue–Stieltjes measure, which may be associated to any function of bounded variation on the real line. The

Lebesgue–Stieltjes measure is a regular Borel measure, and conversely every regular Borel measure on the real line is of this kind.

Lebesgue–Stieltjes integrals, named for Henri Leon Lebesgue and Thomas Joannes Stieltjes, are also known as Lebesgue–Radon integrals or just Radon integrals, after Johann Radon, to whom much of the theory is due. They find common application in probability and stochastic processes, and in certain branches of analysis including potential theory.

Definition

The Lebesgue–Stieltjes integral

$$\int_a^b f(x)dg(x)$$

is defined when $f : [a,b] \to R$ is Borel-measurable and bounded and $g : [a,b] \to R$ is of bounded variation in $[a,b]$ and right-continuous, or when f is non-negative and g is monotone and right-continuous. To start, assume that f is non-negative and g is monotone non-decreasing and right-conti-nuous. Define $w((s,t]) := g(t) - g(s)$ and $w(\{a\})$: $= 0$ (Alternatively, the cons-truction works for g left-continuous, $w([s,t))$: $= g(t) - g(s)$ and $w(\{b\}) := 0$).

By Carathéodory's extension theorem, there is a unique Borel measure μ_g on $[a,b]$ which agrees with w on every interval I. The measure μ_g arises from an outer measure (in fact, a metric outer measure) given by

$$\mu_g(E) = \inf\{\sum_i \mu_g(I_i) \mid E \subset \bigcup_i I_i\}$$

the infimum taken over all coverings of E by countably many semiopen intervals. This measure is sometimes called the Lebesgue–Stieltjes measure associated with g.

The Lebesgue–Stieltjes integral

$$\int_a^b f(x)dg(x)$$

is defined as the Lebesgue integral of f with respect to the measure μ_g in the usual way. If g is non-increasing, then define

$$\int_a^b f(x)dg(x) := -\int_a^b f(x)d(-g)(x),$$

the latter integral being defined by the preceding construction. If g is of bounded variation and f is bounded, then it is possible to write

$$g(x) = g_1(x) - g_2(x)$$

where $g_1(x) := Vxag$ is the total variation of g in the interval $[a,x]$, and $g_2(x) = g_1(x) - g(x)$. Both g_1 and g_2 are monotone non-decreasing. Now the Lebesgue–Stieltjes integral with respect to g is defined by

$$\int_a^b f(x)dg(x) = \int_a^b f(x)dg_1(x) - \int_a^b f(x)dg_2(x),$$

where the latter two integrals are well-defined by the preceding construction.

Daniell Integral

An alternative approach (Hewitt & Stromberg 1965) is to define the Lebesgue–Stieltjes integral as the Daniell integral that extends the usual Riemann–Stieltjes integral. Let g be a non-increasing right-continuous function on $[a,b]$, and define $I(f)$ to be the Riemann–Stieltjes integral

$$I(f) = \int_a^b f(x)dg(x)$$

for all continuous functions f. The functional I defines a Radon measure on $[a,b]$. This functional can then be extended to the class of all non-negative functions by setting

$$\overline{I}(h) = \sup\{I(f) \mid f \in C[a,b], 0 \le f \le h\}$$

and

$$\overline{\overline{I}}(h) = \inf\{I(f) \mid f \in C[a,b], h \le f\}.$$

For Borel measurable functions, one has

$$\overline{I}(h) = \overline{\overline{I}}(h),$$

and either side of the identity then defines the Lebesgue–Stieltjes integral of h. The outer measure μ_g is defined via

$$\mu_g(A) = \overline{\overline{I}}(\chi_A)$$

where χ_A is the indicator function of A. Integrators of bounded variation are handled as above by decomposing into positive and negative variations.

Example

Suppose that $\gamma:[a,b]\to\mathbb{R}^2$ is a rectifiable curve in the plane and $\rho:\mathbb{R}^2\to[0,\infty)$ is Borel measurable. Then we may define the length of γ with respect to the Euclidean metric weighted by ρ to be $\int_a^b \rho(\gamma(t))d\ell(t)$, where $\ell(t)$ is the length of the restriction of γto $[a,t]$. This is sometimes called the ρ-length of γ. This notion is quite useful for various applications: for example, in muddy terrain the speed in which a person can move may depend on how deep the mud is. If $\rho(z)$ denotes the inverse of the walking speed at or near z, then the ρ-length of γ is the time it would take to traverse γ. The concept of extremal length uses this notion of the ρ-length of curves and is useful in the study of conformal mappings.

Integration by Parts

A function f is said to be "regular" at a point if the right $f(a+)$ and left hand limits and $f(a-)$ exist, and the function takes the average value,

$$f(a)=\frac{1}{2}\big(f(a-)+f(a+)\big),$$

at the limiting point. Given two functions U and V of finite variation, if at each point either U or V is continuous, or if both U and V are regular, then there is an integration by parts formula for the Lebesgue–Stieltjes integral:

$$\int_a^b U\,dV+\int_a^b V\,dU=U(b+)V(b+)-U(a-)V(a-),$$

where $b>a$. Under a slight generalization of this formula, the extra conditions on U and V can be dropped.

An alternative result, of significant importance in the theory of Stochastic calculus is the following. Given two functions U and V of finite variation, which are both right-continuous and have left-limits (they are cadlag functions) then

$$U(t)V(t)=U(0)V(0)+\int_{(0,t]}U\,dV+\int_{(0,t]}V\,dU+\sum\nolimits_{u\in(0,t]}\Delta U_u\Delta V_u,$$

where $\Delta U_u=U(t)-U(t-)$.

This result can be seen as a precursor to Ito's lemma, and is of use in the general theory of Stochastic integration. The final term is $\Delta U(t)\Delta V(t) = d[U,V]$, which arises from the quadratic covariation of *U*and *V*.

Related Concepts

Lebesgue Integration: When $g(x) = x$ for all real x, then μ_g is the Lebesgue measure, and the Lebesgue–Stieltjes integral of f with respect to g is equivalent to the Lebesgue integral of f.

Riemann–Stieltjes Integration and Probability Theory

Where f is a continuous real-valued function of a real variable and v is a non-decreasing real function, the Lebesgue–Stieltjes integral is equivalent to the Riemann–Stieltjes integral, in which case we often write

$$\int_a^b f(x)dv(x)$$

for the Lebesgue–Stieltjes integral, letting the measure μ_v remain implicit. This is particularly common in probability theory when v is the cumulative distribution function of a real-valued random variable X, in which case

$$\int_{-\infty}^{\infty} f(x)dv(x) = \mathrm{E}[f(X)].$$

Daniell Integral

In mathematics, the Daniell integral is a type of integration that generalizes the concept of more elementary versions such as the Riemann integral to which students are typically first introduced. One of the main difficulties with the traditional formulation of the Lebesgue integral is that it requires the initial development of a workable measure theory before any useful results for the integral can be obtained.

However, an alternative approach is available, developed by Percy J. Daniell (1918) that does not suffer from this deficiency, and has a few significant advantages over the traditional formulation, especially as the integral is generalized into higher dimensional spaces and further generalizations such as the Stieltjes integral. The basic idea involves the axiomatization of the integral.

The Daniell Axioms

We start by choosing a family H of bounded real functions (called *elementary functions*) defined over some set X, that satisfies these two axioms:

1. H is a linear space with the usual operations of addition and scalar multiplication.
2. If a function $h(x)$ is in H, so is its absolute value $|h(x)|$.

In addition, every function h in H is assigned a real number Ih, which is called the *elementary integral* of h, satisfying these three axioms:

1. Linearity. If h and k are both in H, αand βand are any two real numbers, then $I(\alpha h+\beta k)=\alpha Ih+\beta Ik$.
2. Nonnegativity. If $h(x)\geq 0$, then $Ih\geq 0$.
3. Continuity. If $h_n(x)$ is a nonincreasing sequence (i.e. $h_1\geq\cdots\geq h_k\geq\cdots$) of functions in H that converges to 0 for all x in X, then $Ih_n\rightarrow 0$.

That is, we define a continuous non-negative linear functional over the space of elementary functions.

These elementary functions and their elementary integrals may be any set of functions and definitions of integrals over these functions which satisfy these axioms. The family of all step functions evidently satisfies the above axioms for elementary functions. Defining the elementary integral of the family of step functions as the (signed) area underneath a step function evidently satisfies the given axioms for an elementary integral.

Applying the construction of the Daniell integral described further below using step functions as elementary functions produces a definition of an integral equivalent to the Lebesgue integral. Using the family of all continuous functions as the elementary functions and the traditional Riemann integral as the elementary integral is also possible, however, this will yield an integral that is also equivalent to Lebesgue's definition.

Doing the same, but using the Riemann–Stieltjes integral, along with an appropriate function of bounded variation, gives a definition of integral equivalent to the Lebesgue–Stieltjes integral. Sets of measure zero may be defined in terms of elementary functions as follows. A set Z which is a subset of X is a set of measure zero if for any $\epsilon > 0$, there exists a nondecreasing sequence of nonnegative elementary functions $h_p(x)$ in H such that $Ih_p < \epsilon$ and $\sup_p h_p(x) \geq 1$ on Z.

A set is called a set of full measure if its complement, relative to X, is a set of measure zero. We say that if some property holds at every point of a set of full measure (or equivalently everywhere except on a set of measure zero), it holds almost everywhere.

Definition of the Daniell Integral

We can then proceed to define a larger class of functions, based on our chosen elementary functions, the class L^+, which is the family of all functions that are the limit of a nondecreasing sequence h_n of elementary functions almost everywhere, such that the set of integrals Ih_n is bounded. The integral of a function f in L^+ is defined as:

$$If = \lim_{n\to\infty} Ih_n$$

It can be shown that this definition of the integral is well-defined, i.e. it does not depend on the choice of sequence h_n. However, the class L^+ is in general not closed under subtraction and scalar multiplication by negative numbers, but we can further extend it by defining a wider class of functions L such that every function $\phi(x)$ can be represented on a set of full measure as the difference $\phi = f - g$, for some functions f and g in the class L^+. Then the integral of a function $\phi(x)$ can be defined as:

$$\int_X \phi(x)dx = If - Ig$$

Again, it may be shown that this integral is well-defined, i.e. it does not depend on the decomposition of into f and g. This is the final construction of the Daniell integral.

Properties

Nearly all of the important theorems in the traditional theory of the Lebesgue integral, such as Lebesgue's dominated convergence theorem, the Riesz–Fischer theorem, Fatou's lemma, and Fubini's theorem may also readily be proved using this construction. Its properties are identical to the traditional Lebesgue integral.

Measures from the Daniell integral: Because of the natural correspondence between sets and functions, it is also possible to use the Daniell integral to construct a measure theory. If we take the characteristic function $\chi(x)$ of some set, then its integral may be taken as the measure of the set. This definition of measure based on the Daniell integral can be shown to be equivalent to the traditional Lebesgue measure.

Advantages Over the Traditional Formulation

This method of constructing the general integral has a few advantages over the traditional method of Lebesgue, particularly in the field of functional analysis. The Lebesgue and Daniell constructions are equivalent, as pointed out above, if ordinary finite-valued step functions are chosen as elementary functions. However, as one tries to extend the definition of the integral into more complex domains (e.g. attempting to define the integral of a linear functional), one runs into practical difficulties using Lebesgue's construction that are alleviated with the Daniell approach. The Polish mathematician Jan Mikusinski has made an alternative and more natural formulation of Daniell integration by using the notion of absolutely convergent series. His formulation works for Bochner integral (Lebesgue integral for mappings taking values in Banach spaces).

Mikusinski's lemma allows one to define integral without mentioning null sets. He also proved change of variables theorem for multiple integral for Bochner integrals and Fubini's theorem for Bochner integrals using Daniell integration. The book by Asplund and Bungart carries a lucid treatment of this approach for real valued functions. It also offers a proof of an abstract Radon–Nikodym theorem using Daniell–Mikusinski approach.

Haar Integral

Using the general theory of Lebesgue integration, one can then define an integral for all Borel measurable functions f on G. This integral is called the Haar integral. If μ is a left Haar measure, then

$$\int_G f(sx)\, d\mu(x) = \int_G f(x)\, d\mu(x)$$

for any integrable function f. This is immediate for characteristic functions, being essentially the definition of left invariance.

Examples

- A Haar measure on the topological group (R, +) which takes the value 1 on the interval [0,1] is equal to the restriction of Lebesgue measure to the Borel subsets of R. This can be generalized to (R^n, +).
- If G is the group of nonzero real numbers with multiplication as operation, then a Haar measure μ is given by

$$\mu(S) = \int_S \frac{1}{|t|} dt$$

for any Borel subset S of the nonzero reals.

This generalizes to the following:

For $G = GL(n,R)$, any left Haar measure is a right Haar measure and one such measure μ is given by

$$\mu(S) = \int_S \frac{1}{|\det(X)|^n} dX$$

where dX denotes the Lebesgue measure on $\mathbb{R}n^2$, the set of all $n \times n$-matrices. This follows from the change of variables formula.

More generally, on any Lie group of dimension d a left Haar measure can be associated with any non-zero left-invariant d-form ω, as the *Lebesgue measure* $|\omega|$; and similarly for right Haar measures.

This means also that the modular function can be computed, as the absolute value of the determinant of the adjoint representation. In order to define a Haar measure μ on the unit circle T, consider the function f from $[0,2\pi]$ onto T defined by $f(t) = (\cos(t),\sin(t))$. Then μ can be defined by

$$\mu(S) = \frac{1}{2\pi} m\left(f^{-1}(S)\right),$$

where m is the Lebesgue measure. The factor $(2\pi)^{-1}$ is chosen so that $\mu(T) = 1$.

If G is the group of non-null quaternions, then G can be seen as an open subset of R^4. A Haar measure μ is given by

$$\mu(S) = \int_S \frac{1}{(x^2 + y^2 + z^2 + w^2)^2} dx\,dy\,dz\,dw$$

where $dx\ dy\ dz\ dw$ denotes the Lebesgue measure in R^4 and S is a Borel subset of G.

Henstock–Kurzweil Integral

In mathematics, the Henstock–Kurzweil integral, also known as the Denjoy integral (pronounced [dQÞÈ'wa]) and the Perron integral, is one of a number of definitions of the integral of a function. It is a generalization of the Riemann integral which in some situations is more useful than the Lebesgue integral.

This integral was first defined by Arnaud Denjoy (1912). Denjoy was interested in a definition that would allow one to integrate functions like

$$f(x) = \frac{1}{x}\sin\left(\frac{1}{x^3}\right).$$

This function has a singularity at 0, and is not Lebesgue integrable. However, it seems natural to calculate its integral except over $[-\varepsilon,\delta]$ and then let $\varepsilon,\ \delta \to 0$.

Trying to create a general theory Denjoy used transfinite induction over the possible types of singularities which made the definition quite complicated. Other definitions were given by Nikolai Luzin (using variations on the notions of absolute continuity), and by Oskar Perron, who was interested in continuous major and minor functions. It took a while to understand that the Perron and Denjoy integrals are actually identical.

Later, in 1957, the Czech mathematician Jaroslav Kurzweil discovered a new definition of this integral elegantly similar in nature to Riemann's original definition which he named the gauge integral; the theory was developed by Ralph Henstock. Due to these two important mathematicians, it is now commonly known as the Henstock–Kurzweil integral. The simplicity of Kurzweil's definition made some educators advocate that this integral should replace the Riemann integral in introductory calculus courses, but this idea has not gained traction.

Definition

Henstock's definition is as follows:

Given a tagged partition P of $[a, b]$, say

$$a = u_0 < u_1 < \cdots < u_n = b,\ t_i \in [u_{i-1}, u_i]$$

and a positive function

$$\delta : [a, b] \to (0, \infty),$$

which we call a *gauge*, we say P is δ-fine if

$$\forall i\ \ t_i - \delta(t_i) < u_{i-1} \le t_i \le u_i < t_i + \delta(t_i).$$

For a tagged partition P and a function

$$f : [a, b] \to \mathbb{R}$$

we define the Riemann sum to be

$$\sum_P f = \sum_{i=1}^{n} (u_i - u_{i-1}) f(t_i).$$

Given a function

$$f : [a, b] \to \mathbb{R},$$

we now define a number I to be the Henstock–Kurzweil integral of f if for every $\varepsilon > 0$ there exists a gauge δ such that whenever P is δ-fine, we have

$$\left| \sum_P f - I \right| < \varepsilon.$$

If such an I exists, we say that f is Henstock–Kurzweil integrable on $[a, b]$. Cousin's theorem states that for every gauge δ, such a δ-fine partition P does exist, so this condition cannot be satisfied vacuously. The Riemann integral can be regarded as the special case where we only allow constant gauges.

Properties

Let f: $[a, b] \to$ R be any function. If $a < c < b$, then f is Henstock–Kurzweil integrable on $[a, b]$ if and only if it is

Henstock–Kurzweil integrable on both $[a, c]$ and $[c, b]$, and then

$$\int_a^b f(x)dx = \int_a^c f(x)dx + \int_c^b f(x)dx.$$

The Henstock–Kurzweil integral is linear, i.e., if f and g are integrable, and α, β are reals, then $\alpha f + \beta g$ is integrable and

$$\int_a^b \alpha f(x) + \beta g(x)dx = \alpha\int_a^b f(x)dx + \beta\int_a^b g(x)dx.$$

If f is Riemann or Lebesgue integrable, then it is also Henstock–Kurzweil integrable, and the values of the integrals are the same. The important Hake's theorem states that

$$\int_a^b f(x)dx = \lim_{c\to b-}\int_a^c f(x)dx$$

whenever either side of the equation exists, and symmetrically for the lower integration bound. This means that if f is "improperly Henstock–Kurzweil integrable", then it is properly Henstock–Kurzweil integrable; in particular, improper Riemann or Lebesgue integrals such as

$$\int_0^1 \frac{\sin(1/x)}{x}dx$$

are also Henstock–Kurzweil integrals. This shows that there is no sense in studying an "improper Henstock–Kurzweil integral" with finite bounds. However, it makes sense to consider improper Henstock–Kurzweil integrals with infinite bounds such as

$$\int_a^{+\infty} f(x)dx := \lim_{b\to+\infty}\int_a^b f(x)dx.$$

For many types of functions the Henstock–Kurzweil integral is no more general than Lebesgue integral. For example, if f is bounded, the following are equivalent:

- f is Henstock–Kurzweil integrable,
- f is Lebesgue integrable,
- f is Lebesgue measurable.

In general, every Henstock–Kurzweil integrable function is measurable, and f is Lebesgue integrable if and only if both

f and $|f|$ are Henstock–Kurzweil integrable. This means that the Henstock–Kurzweil integral can be thought of as a "non-absolutely convergent version of Lebesgue integral". It also implies that the Henstock–Kurzweil integral satisfies appropriate versions of the monotone convergence theorem (without requiring the functions to be nonnegative) and dominated convergence theorem (where the condition of dominance is loosened to $g(x) \leq f_n(x) \leq h(x)$ for some integrable g, h).

If F is differentiable everywhere (or with countable many exceptions), the derivative F2† is Henstock–Kurzweil integrable, and its indefinite Henstock–Kurzweil integral is F. (Note that F2† need not be Lebesgue integrable.) In other words, we obtain a simpler and more satisfactory version of the second fundamental theorem of calculus: each differentiable function is, up to a constant, the integral of its derivative:

$$F(x) - F(a) = \int_a^x F'(t)dt.$$

Conversely, the Lebesgue differentiation theorem continues to holds for the Henstock–Kurzweil integral: if f is Henstock–Kurzweil integrable on $[a, b]$, and

$$F(x) = \int_a^x f(t)dt,$$

then $F'(x) = f(x)$ almost everywhere in $[a, b]$ (in particular, F is almost everywhere differentiable). The space of all Henstock–Kurzweil-integrable functions is often endowed with the Alexiewicz norm, with respect to which it is barrelled but incomplete.

Chapter 2

Derivative Equations

Rate of Change

The derivative of a function at a chosen input value describes the best linear approximation of the function near that input value. For a real-valued function of a single real variable, the derivative at a point equals the slope of the tangent line to the graph of the function at that point. In higher dimensions, the derivative of a function at a point is a linear transformation called the linearization. The process of finding a derivative is called differentiation.

The fundamental theorem of calculus states that differentiation is the reverse process to integration. The mathematical relationship between an independent and dependent dimension. The relationship expressed in the form of a mathematical function defines a situation for a given set of conditions and properties. Since we allowed the independent dimension to change, it was no longer a fixed quantity but became a variable.

The graph of the function reflected a horizontal change of the independent variable with a vertical change in the dependent dimension.

This graph allows us to visualize how a situation changes with respect to a change in the conditions. To find out how much the function, $f(x)$, changes between two points x_1 and x_2, we simply enter in the two values for the independent variable x and then calculate the difference between the dependent variable, f, for those given conditions. Remember that a variable is nothing more than a dimension that is allowed to change or take on any value. Thus, from x_1 to x_2, the change in the independent variable, referred to as Δx is:

$$\Delta x = x_1 - x_2$$

The corresponding change in the dependent variable , referred to as Δf is:

$\Delta f = f(x_2) - f(x_1)$ Δx refers to an interval over which we are analyzing the change in the dependent dimension, f.. The second point x_2 can be written in terms of the first point x_1 plus the change in the variable, Δx

$$x_2 - x_1 = \Delta x$$

$$\therefore \quad x_2 = x_1 + \Delta x$$

Therefore the change in the dependent variable over an interval $x_1 x_2$ from $\Delta f = f(x_2) - f(x_1)$ which can also be written as:

$$\Delta f = f(x_1 + \Delta x) - f(x_1)$$

where $\Delta x = x_2 - x_1$. For example the change in the function $f(x) = 3 \cdot x^3$ $x = 3$ to $x = 5$, where Δx is 5 – 3 = 2 is:

$$\Delta f = f(x_1 + \Delta x) - f(x_1)$$

$$\Delta f = f(3+2) - f(3)$$

$$\Delta f = 3.5^3 - 3.3^3 = 375 - 81$$

$$\Delta f = 294$$

Average Rate of Change

We have learned that a change in the independent variable is defined as $\Delta x = x_2 - x_1$, and the corresponding change in the dependent variable over this interval is $\Delta f = f(x_1 + \Delta x) - f(x_1)$. The question we now must ask ourselves is how can we measure the relative change of the dependent variable with respect to the independent variable? In other words how can we calculate how much more or less Δf changed compared to Δx.

To calculate how much more $f(x)$ changed over an interval from x_1, x_2, we simply divide the change in f over the change in x for the interval. Thus we divide, Δf by the interval over we which are evaluating it, Δx which is equal to $x_2 - x_1$. Thus the relative change of f with respect to x over an interval Δx is defined as:

$$\frac{\Delta f}{\Delta x} = \frac{f(x_2) - f(x_1)}{x_2 - x_1}$$

While this expression may seem rather simple, it does require some explanation. By dividing the change in f by the change in x what we are doing is calculating how much more f changed for a given change in x. For example in the function, $f(x) = 3.x^3$, when x changed from 3 to 5, f changed from 81 to 375. Over this interval of from $x = 3$ to $x = 5$, the Δf was 294.

Thus the relative change in f with respect to a change in the independent variable x is:

$$\frac{\Delta f}{\Delta x} = \frac{f(x_2) - f(x_1)}{x_2 - x_1}$$

The value of 147 tells us that f changes 147 times faster than x over that interval of Δx from $x = 3$ to $x = 5$ only. Thus for each unit change in x, $\Delta x = 1$, the corresponding change in f is 147. We can therefore define the rate of change of a function with respect to its independent variable to be:

$$\frac{\Delta f}{\Delta x} = \frac{f(x_1 + \Delta x) - f(x_1)}{\Delta x}$$

The value, called the rate of change of the function, refers to how much more or less $f(x)$ changes for a unit change in x. Itis only valid is for the interval under consideration, $\Delta x = x_2 - x_1$.

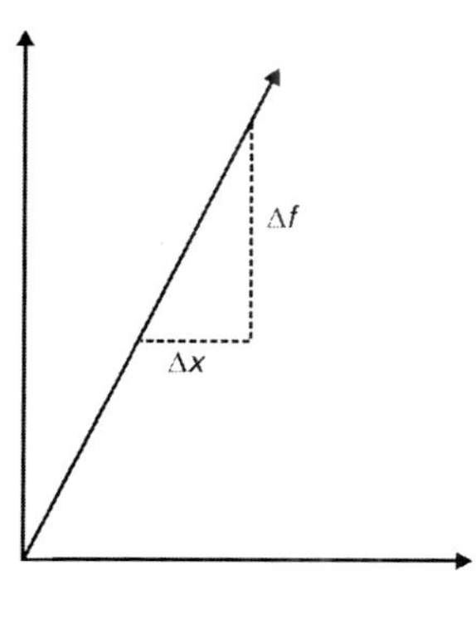

Another way of understanding what rate of change of a function means is to look at the steepness of the line connecting the two endpoints of the interval under consideration.

As Δf increases, the steepness of the line connecting the two endpoints will increase. Thus, the greater the rate of change of the function, the greater its slope or steepness over the interval under consideration.

Since slope and rate of change are synonymous, then how is rate of change defined for functions whose graphs do not have constant slopes? For example, from $x = 9$ to $x = 12$ of $f(x) = 3.x^3$, the change in f is 2997. Thus the rate of change of the function over the interval is:

$$\frac{\Delta f}{\Delta x} = \frac{2997}{3} = 999$$

This value is significantly higher than the rate of change calculated for the previous interval from $x = 3$ to $x = 5$. We can only conclude that the rate of change or slope of the graph must be increasing and is not constant over an interval $\Delta x = x_2 - x_1$. Look at the graph of the function $f(x) = 3.x^3$ to understand how this might be so:

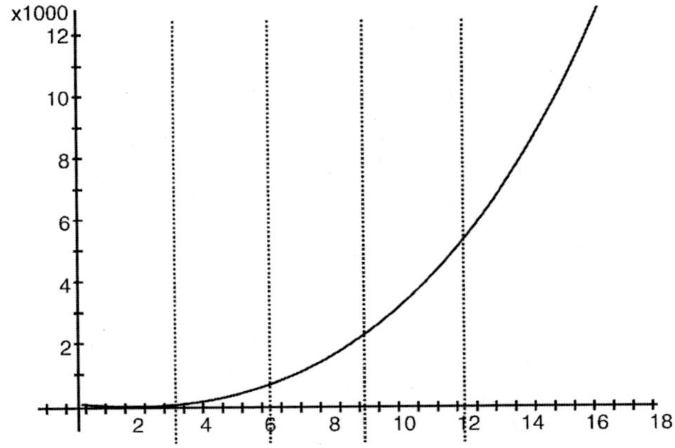

Since the rate of change of a function can change, then we have to come up with a more refined definition of rate of change. We can define the average rate of change of a function over an interval $\Delta x - x_2 - x_1$, to be equal to

$$\text{Average rate of change} = \frac{\Delta f}{\Delta x} = \frac{f(x_1 + \Delta x) - f(x_1)}{\Delta x}$$

In the next section we will take a closer look at how we can define the exact or instantaneous rate of change of a function.

Instantaneous Rate of Change

In this section we will take a much closer look at rate of change and see how we can define the instantaneous rate of change of a function at any point or value of the independent variable.

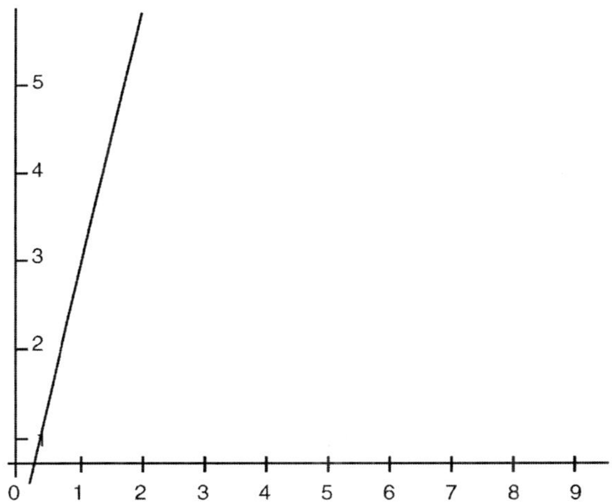

Let us begin with a study of the simple $f(x) = 3x$. Functions of the form $f(x) = cx$, where c is a constant, express direct relationships. This is because the value of the function, *f*, is a constant multiple or fraction of the independent variable.

Thus f is said to be directly proportional to x. The graph of the function $f(x) = 3x$ looks like. From $x = 1$ to $x = 3$, $\Delta x = 2$, and $\Delta f = 9 - 3 = 6$. Thus, over this interval of $\Delta x = 2$, Δf equals 6. The average rate at which f changed with respect to x is by definition, $\frac{\Delta f}{\Delta x} = \frac{6}{2} = 3$. For each unit change in x, the change in f is 3. This tells us that f is changing three times faster that x is changing over the interval from $x = 1$ to $x = 3$. We can now look at the interval from $x = 2$ to $x = 4$ where Δx equals 2.

$$\Delta f = f(x_1 + \Delta x) - f(x_1)$$
$$\Delta f = f(2+2) - f(2)$$
$$\Delta f = 3.(4) - 3.(2)$$
$$\Delta f = 6$$

Thus

$$\frac{\Delta f}{\Delta x} = \frac{f(x_1 + \Delta x) - f(x_1)}{\Delta x} = \frac{6}{2} = 3$$

Over this interval, the rate of change is the same constant, 3. This leads us to conclude that the rate of change of the function over any interval is a constant, 3. This can be proven by the definition of rate of change:

$$f(x) = 3x$$
$$f(x + \Delta x) = 3(x + \Delta x)$$
$$\frac{\Delta f}{\Delta x} = \frac{f(x + \Delta x) - f(x)}{\Delta x} = \frac{3\cancel{x} + 3\Delta x - 3\cancel{x}}{\Delta x}$$
$$\frac{\Delta f}{\Delta x} = \frac{3\Delta\cancel{x}}{\Delta\cancel{x}} = 3$$

The rate of change of the direct function $f(x) = cx$ is c and is constant over an interval, Dx. The graph of a function $f(x) = cx$ is therefore an increasing straight line with a constant slope or steepness equal to c. For this reason functions of the form $f(x) = cx$ are also called linear functions since their graphs are straight lines with a constant rate of change.

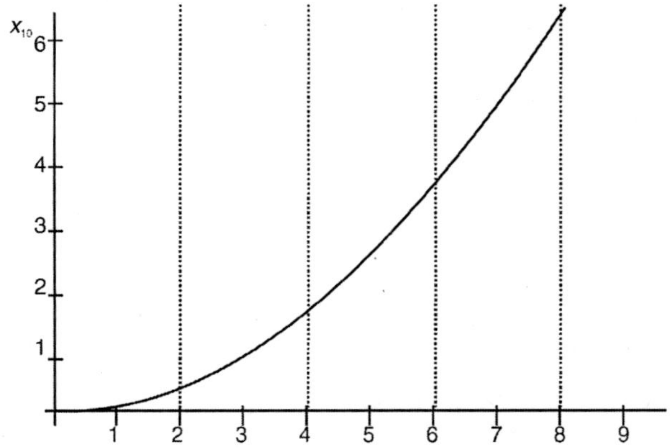

On the other hand, the graph of the function $f(x) = x^2$ is not a straight line. Unlike a line, the rate of change of $f(x)$ is not the same constant over any interval Dx. To define the rate of change for the function we will have to derive a more precise way of defining rate of change of function. To do this we will analyze the function over small intervals of *Dx*

The first point to notice is that the rate of change of the function varies with x. When x is small, f does not change that much as compared to how much it changes when x is large. We can conclude that the rate of change of f with respect to x, is not constant over any interval, *Dx*, but varies with x. To begin our analysis let us divide up the graph into intervals of *Dx = 2* and study what is happening in each of these intervals separately. From $x = 0$ to $x = 2$, Dx = 2, the change in f, *Df, is* $f(2) - f(0) = 4$. From $x = 6$ to $x = 8$, *Dx* = 2, the change in *f, Df, is* $f(8) - f(6) = 28$.

Clearly, as x increases the rate of change of the function is increasing and is not a constant as in our study of the line where the function changed at the same rate as x increased.

Returning back to the graph of $f(x) = x^2$; we calculated that from $x = 0$ *to* $x = 2$ the change in f was 4. We can then conclude that from $x = 0$ to $x = 2$ the average rate of change of the function over that particular interval is 2 or:

$$\frac{\Delta f}{\Delta x} = \frac{4}{2} = 2$$

Remember it is called average because this rate of change is only valid from $x = 0$ to $x = 2$. Now let us consider the interval from $x = 2$ to $x = 4$ where once again the change in x, *D*x, is 2. The change in f of the graph is equal to:

$$\begin{aligned}\Delta f &= f(x+\Delta x) - f(x)\\ &= f(2+2) - f(2)\\ &= f(4) - f(2)\\ &= 16 - 4\\ &= 12\end{aligned}$$

Thus the average rate of change of the function over this interval is equal to

$$\frac{\Delta f}{\Delta x} = \frac{12}{2} = 6$$

This value is greater than the value we observed from $x = 0$ to $x = 2$. This implies that from $x = 2$ to $x = 4$, $f(\mathrm{x})$ is increasing at a greater rate than from $x = 0$ to $x = 2$. This is despite the fact that in both cases the $Dx = 2$. The rate of change is therefore not constant over the but is increasing with x. When we say the rate of change of $f(x)$ from $x = 2$ to $x = 4$ is 6, it is only the average value for the given interval since it assumes rate of change is constant over that interval. Let us move on to the next interval of $x = 4$ to $x = 6$. Once again the change in x or Dx equals 2 but the corresponding change in $f(x)$ is not the same as before.

$$\begin{aligned}\Delta f &= f(x + \Delta x) - f(x)\\ &= (x + \Delta x)^2 - (x)^2\\ &= x^2 + 2x\Delta x + \Delta x^2 - x^2\\ &= 2x\Delta x + \Delta x^2\end{aligned}$$

For x equal to 4 and $Dx = 2$, the change in the function over this interval is

$$\Delta f = 2x\Delta x + \Delta x^2$$

$$\Delta f = 2(4)(2) + (2)^2$$

$$\Delta f = 20$$

Note that this corresponds to the same value we would get by:

$$\begin{aligned}\Delta f &= f(6) - f(4)\\ &= 6^2 - 4^2\\ &= 36 - 16\\ &= 20\end{aligned}$$

The average rate of change over this interval is therefore:

$$\frac{\Delta f}{\Delta x} = \frac{20}{2} = 10$$

This is still larger than the rate of change for the previous interval which was 6. Remember rate of change, by definition, refers to how much the function changes with respect to a change in the independent variable. The steepness or slope of the line over the interval provides a geometric understanding for this concept.

The average rate of changes calculated over each interval can be used to approximate the graph of $f(x)$ from $x = 0$ *to* $x = 8$.

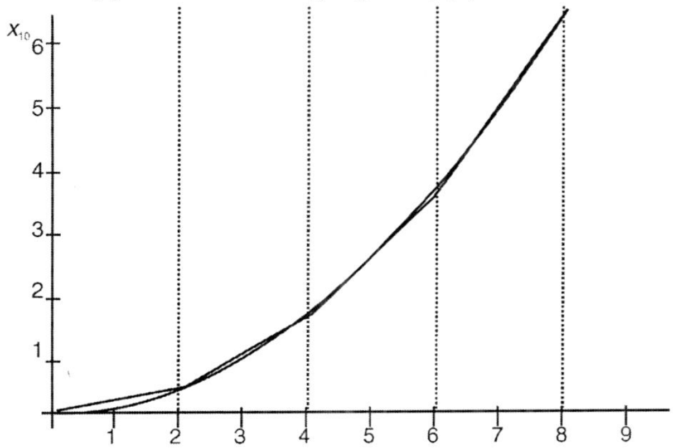

This roughly corresponds to the original graph, but as we can see the rate of change or slope is not constant through out the interval from $x = 0$ *to* $x = 8$, but increases as x-increases. In order to get more accurate answers we need to reduce our interval of $Dx = 2$ to a much smaller one.

The idea being we need to analyze our graph over a small interval, *Dx*, to see what exactly is going on at each instant the function is changing. Here is where we begin our study of Calculus. We break down and freeze a changing situation into an infinite series of actions and analyzing what is going on in each individual actions.

Basic Properties of Derivative

We continue our study of classes of functions which are suitably restricted. Again we are passing from the general to the particular. The next most particular class of function we study after the class of continuous functions is the class of differentiable functions. We discuss the definition, show how to get "new functions from old" in what by now is a fairly routine way, and prove that this is a smaller class: that every differentiable function is continuous, but that there are continuous functions that are not differentiable. Informally, the difference is that the graph of a differentiable function may not have any sharp corners in it.

As with continuous functions, our aim is to show that there are many attractive properties which hold for differentiable functions that don't hold in general for continuous functions. One we discuss in some detail is the ease with which certain limits can be evaluated, by introducing l'Hopital's rule. Although we don't prove this, the corresponding results are false for continuous functions.

We take the view that much of this material has already been discussed last year, so we move fairly quickly over the basics.

Definition: Let U be an open subset of $\mathbb{R}$, and let f : $U \to \mathbb{R}$. We say that f is differentiable at $a \in U$ iff

$$\lim_{x \to \alpha} \frac{f(x) - f(a)}{x - a}$$

or equivalently,

$$\lim_{h \to 0} \frac{f(a+h) - f(a)}{h}$$

exists. The limit, if it exists, is written as $f'(a)$. We say that f is differentiable in U if and only if it is differentiable at each a in U. Note that the Newton quotient is not defined when $x = a$, nor need it be for the definition to make sense. But the Newton quotient, if it exists, can be extended to be a continuous function at a by defining its value at that point to be $f'(a)$. Note also the emphasis on the existence of the limit. Differentiation is as much about showing the *existence* of the derivative, as *calculating* the value of the derivative.

Example: Let $f(x) = x^3$. Show, directly from the definition, that $f'(a) = 3a^2$.

Solution: This is just another way of asking about particular limits, we must compute

$$\lim_{x \to \alpha} \frac{x^3 - a^3}{x - a} = \lim_{x \to \alpha} \frac{(x-a)(x^2 + xa + a^2)}{x - a} = \lim_{x \to \alpha} (x^2 + 2xa + a^2) = 3a^2.$$

Exercise: Let $f(x) = \sqrt{x}$. Show, directly from the definition, that $f'(a) = 1/2\ \sqrt{a}$ when $a \neq 0$. What function do you have to consider in the particular case when $a = 4$?

Just as with continuity, it is impractical to use this definition every time to compute derivatives; we need results showing how to differentiate the usual class of functions, and we assume these are known from last year. Thus we assume the rules for differentiation of sums products and compositions of functions, together with the known derivatives of elementary functions such as sin, cos and tan; their reciprocals sec, *cosec* and cot; and exp and log.

Proposition: Let f and g be differentiable at a, and let k be a constant. Then $k.f$, $f + g$ and fg are differentiable at f. Also, if $g(a) \neq 0$, then f/g is differentiable at a. Let f be differentiable at a, and let g be differentiable at $f(a)$. Then gof is differentiable at a. In addition, the usual rules for calculating these derivatives apply.

Example: Let $f(x) = \tan\left(\frac{x^2 - a^2}{x^2 + a^2}\right)$ for $a \neq 0$. Show that f is differentiable at every point of its domain, and calculate the derivative at each such point.

Solution: This is the same example we considered in. There we showed the domain was the whole of $\mathbb{R}$, and that the function was continuous everywhere. Let $g(x) = \frac{x^2 - a^2}{x^2 + a^2}$. Then g is properly defined for all values of x, and the quotient is differentiable, since each term is, and since $x^2 + a^2 \neq 0$ for any x since $a \neq 0$. Thus f is differentiable using the chain rule since $f =$ tanog, and we are assuming known that the elementary functions like tan are differentiable.

Finally to actually calculate the derivative, we have:

$$f'(x) = sec^2\left(\frac{x^2-a^2}{x^2+a^2}\right)\frac{\left(x^2+a^2\right)\cdot 2x-\left(\left(x^2-a^2\right)\cdot 2x\right)}{\left(x^2+a^2\right)^2}.$$

$$= \frac{4a^2x}{\left(x^2+a^2\right)^2}\cdot \sec^2\left(\frac{x^2-a^2}{x^2+a^2}\right)$$

Exercise: Let $f(x) = \exp\left(\frac{1+x^2}{1-x^2}\right)$. Show that f is differentiable at every point of its domain, and calculate the derivative at each such point. The first point in our study of differentiable functions is that it is *more* restrictive for a function to be differentiable, than simply to be continuous.

Proposition: Let f be differentiable at a. Then f is continuous at a.

Proof: To establish continuity, we must prove that $\lim_{x\to a} f(x) = f(a)$. Since the Newton quotient is known to converge, we have for $x \neq a$,

$$f(x) - f(a) = \frac{f(x)-f(a)}{x-a}.(x-a) \to f'(a).0 \quad \text{as } x \to a.$$

Hence f is continuous at a.

Example: Let $f(x) = |x|$; then f is continuous everywhere, but *not* differentiable at 0.

Solution: We already know from Example that $|x|$ is continuous. We compute the Newton quotient directly; recall that $|x| = x$ if $x \geq 0$, while $|x| = -x$ if $x < 0$. Thus

$$\lim_{x\to 0+}\frac{f(x)-f(0)}{x-0} = \lim_{x\to 0+}\frac{x-0}{x-0}, \text{ while}$$

$$\lim_{x\to 0-}\frac{f(x)-f(0)}{x-0} = \lim_{x\to 0-}\frac{-x-0}{x-0} = -1.$$

Thus both of the one-sided limits exist, but are unequal, so the limit of the Newton quotient does not exist.

Simple Limits

Our calculus of differentiable functions can be used to compute limits which otherwise prove troublesome.

Proposition: Let f and g be functions such that $f(a) = g(a) = 0$ while $f'(a)$ and $g'(a)$ both exits and $g'(a) \neq 0$. Then

$$\lim_{x \to a} \frac{f(x)}{g(x)} = \frac{f'(a)}{g'(a)}.$$

Proof: Since $f(a) = g(a) = 0$, provided $x \neq a$, we have

$$\frac{f(x)}{g(x)} = \frac{f(x) - f(a)}{g(x) - g(a)}$$

$$= \frac{f(x) - f(a)}{x - a} \frac{x - a}{g(x) - g(a)} \to \frac{f'(a)}{g'(a)} \text{ as } x \to a$$

where the last limit exists, since $g'(a) \neq 0$.

Remark: If $f'(a)$ and $g'(a)$ exist, $\lim_{x \to \alpha} \frac{f(x)}{g(x)}$ computing is easy, since f and g must be continuous at a by Proposition, unless we get an indeterminate form 0/0 or ∞ / ∞ for the formal quotient. In fact l'Hopitals rule helps in both cases, although we need to develop stronger forms.

Example: Show that $\lim_{x \to 0} \frac{3x - \sin x}{x} = 2$.

Solution: Note first that we cannot get the result trivially from, since since $g(a) = 0$ and so we get the indeterminate form 0/0. However, we are in a position to apply the simple form of l'Hopital, since $x' = 1 \neq 0$. Applying the rule gives

$$\lim_{x \to 0} \frac{3x - \sin x}{x} = \lim_{x \to 0} \frac{3 - \cos x}{1} = 2.$$

Example: Show that $\lim_{x \to 0} \frac{\sqrt{1 + x} - 1}{x} = 1/2$.

Solution: Note again that we cannot get the result trivially from 4.16, since this gives the indeterminate 0/0 form, because $g(a) = 0$. However, we are in a position to apply the simple form of l'Hopital, since $x' = 1 \neq 0$. Applying the rule gives

$$\lim_{x \to 0} \frac{\sqrt{1+x}-1}{x} = \lim_{x \to 0} \frac{2^{-1}(1+x)^{-1/2}}{1} = 1/2.$$

Example: Show that $\lim_{x \to 0} \frac{\sin x}{x} = 1.$

Solution: This is spurious because we need the limit to calculate the derivative in the first place, but applying l'Hopital certainly gives the result.

Rolle and the Mean Value Theorem

We can combine our definition of derivative with the Intermediate Value Theorem to give a useful result which is in fact the basis of most elementary applications of the differential calculus. Like the results on continuous functions, it is a global result, and so needs continuity and differentiability on a whole interval.

Theorem (Rolle's Theorem): Let f be continuous on $[a, b]$, and differentiable on (a, b), and suppose that $f(a) = f(b)$. Then there is some c with $a < c < b$ such that $f'(c) = 0$.

Note: The theorem guarantees that the point c exists somewhere. It gives *no* indication of how to find c. Here is the diagram to make the point geometrically:

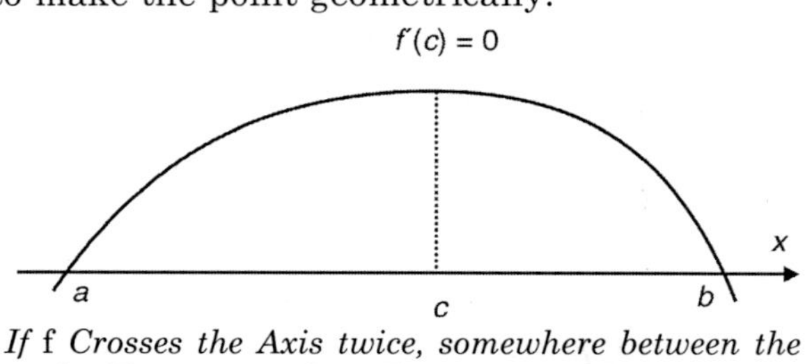

Figure: *If* f *Crosses the Axis twice, somewhere between the two Crossings, the Function is flat. The Accurate Statement of this "Obvious" Observation is Rolle's Theorem.*

Proof: Since f is continuous on the compact interval $[a, b]$, it has both a global maximum and a global minimum. Assume first that the global maximum occurs at an interior point $c \in (a, b)$. In what follows, we pick h small enough so that $c + h$ always lies in (a, b). Then if

$$h > 0, \; \frac{f(c+h)-f(c)}{h} \leq 0, \text{ and so}$$

$$\lim_{h\to 0+} \frac{f(c+h)-f(c)}{h} \leq 0,$$

since we know the limit exists.

Similarly, if $h < 0, \dfrac{f(c+h)-f(c)}{h} \geq 0,$

and so

$$\lim_{h\to 0+} \frac{f(c+h)-f(c)}{h} \geq 0,$$

since we know the limit exists. Combining these, we see that $f'(c) = 0$, and we have the result in this case. A similar argument applies if, instead, the global minimum occurs at the interior point c.

The remaining situation occurs if both the global maximum and global minimum occur at end points; since $f(a) = f(b)$, it follows that f is constant, and any $c \in (a, b)$ will do.

Example: Investigate the number of roots of each of the polynomials

$$p(x) = x^3 + 3x + 1 \quad \text{and} \quad q(x) = x^3 - 3x + 1.$$

Solution: Since $p'(x) = 3(x^2 + 1) > 0$ for all $x \in \mathbb{R}$, we see that p has at most one root; for if it had two (or more) roots there would be a root of $p'(x) = 0$ between them by Rolle. Since $p(0) = 1$, while $p(-1) = -3$, there is at least one root by the Intermediate Value Theorem.

Hence p has exactly one root. We have $q'(x) = 3(x^2 - 1) = 0$ when $x = \pm 1$. Since $q(-1) = 3$ and $q(1) = -1$, there is a root of q between -1 and 1 by the Intermediate Value Theorem. Looking as $x \to \infty$ and as $x \to -\infty$ shows here are three roots of q.

The Mean Value Theorem

Theorem: Let f be continuous on $[a, b]$, and differentiable on (a, b). Then there is some c with $a < c < b$ such that

$$\frac{f(b)-f(a)}{b-a} = f'(c)$$

or equivalently $f(b) = f(a) + (b - a)f'(c)$.

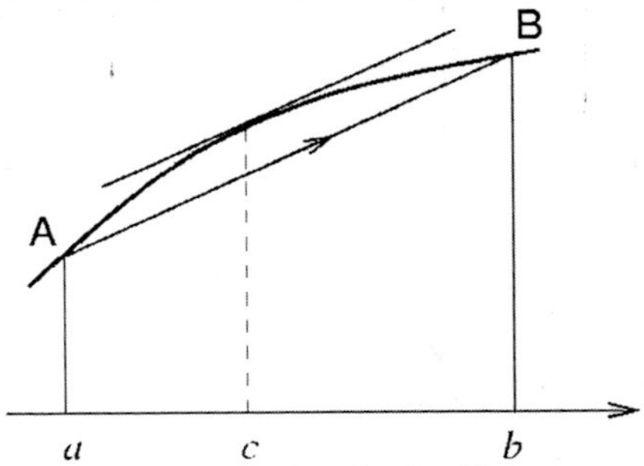

Figure: *Somewhere Inside a chord, the Tangent to* f *will be Parallel to the chord.*

The Accurate Statement of this Common-sense Observation is the Mean Value Theorem.

***Proof*:** We apply Rolle to a suitable function; let

$$h(x) = f(b) - f(x) - \frac{f(b)-f(a)}{b-a}(b - x).$$

Then h is continuous on the interval $[a, b]$, since f is, and in the same way, it is differentiable on the open interval (a, b). Also, $f(b) = 0$ and $f(a) = 0$. We can thus apply Rolle's theorem to h to deduce there is some point c with $a < c < b$ such that $h'(c) = 0$. Thus we have

$$0 = h'(c) = -f'(c) + \frac{f(b)-f(a)}{b-a},$$

which is the required result.

Example: The function f satisfies $f'(x) = \frac{1}{5-x^2}$ and $f(0) = 2$. Use the Mean Value theorem to estimate f.

Solution: We first estimate the given derivative for values of x satisfying $0 < x < 1$. Since for such x, we have $0 < x^2 < 1$, and so $4 < 5 - x^2 < 5$. Inverting we see that

$$\frac{1}{5} < f'(x) < \frac{1}{4} \text{ when } 0 < x < 1.$$

Now apply the Mean Value theorem to f on the interval [0, 1] to obtain some c with $0 < c < 1$ such that $f(1) - f(0) = f'(c)$. From the given value of $f(0)$, we see that $2.2 < f(1) < 2.25$

Exercise: The function f satisfies $f¢(x) \frac{1}{5+\sin x}$ = and $f(0) = 0$. Use the Mean Value theorem to estimate $f(p/2)$. Note the "common sense" description of what we have done. If the derivative doesn't change much, the function will behave linearly. Note also that this gives meaning to the approximation

$$f(a + h) \approx f(a) + hf'(a).$$

We now see that the accurate version of this replaces $f'(a)$ by $f'(c)$ for some c between a and $a + h$.

The Cauchy Mean Value Theorem

Theorem: Let f and g be both continuous on $[a, b]$ and differentiable on (a, b). Then there is some point c with $a < c < b$ such that

$$g'(c)(f(b) - f(a)) = f'(c)(g(b) - g(a)).$$

In particular, whenever the quotients make sense, we have

$$\frac{f(b)-f(a)}{g(b)-g(a)} = \frac{f'(c)}{g'(c)}$$

Proof. Let $h(x) = f(x)(g(b) - g(a)) - g(x)(f(b) - f(a))$, and apply Rolle's theorem exactly as we did for the Mean Value Theorem. Note first that since both f and g are continuous on $[a, b]$, and differentiable on (a, b), it follows that h has these properties.

Also $h(a) = f(a)g(b) - g(a)f(b) = h(b)$. Thus we may apply Rolle to h, to deduce there is some point c with $a < c < b$ such that $h'(c) = 0$. But

$$h'(c) = f'(c)(g(b) - g(a)) - g'(c)(f(b) - f(a))$$

Thus

$$f'(c)(g(b) - g(a)) = g'(c)(f(b) - f(a))$$

This is one form of the Cauchy Mean Value Theorem for f and g. If $g'(c) \neq 0$ for any possible c, then the Mean Value theorem shows that $g(b) - g(a) \neq 0$, and so we can divide the above result to get

$$\frac{f(b)-f(a)}{g(b)-g(a)}=\frac{f'(c)}{g'(c)},$$

giving a second form of the result.

Note: Taking $g(x) = x$ recovers the Mean Value Theorem.

l'Hopital Revisited

We can get a much more useful form of l'Hopital's rule using the Cauchy Mean Value Theorem, directly from the definition of the derivative.

l'Hopital's Rule: General form

Proposition: Let f and g be functions such that $f(a) = g(a) = 0$, and suppose that f and g are differentiable on an open interval I containing a, and that $g'(x) \neq 0$, except perhaps at a. Then

$$\lim_{x\to\alpha}\frac{f(x)}{g(x)}=\lim_{x\to\alpha}\frac{f'(x)}{g'(x)},$$

provided the second limit exists.

Proof: Pick $x > a$ and apply the Cauchy Mean Value Theorem to the interval $[a, x]$, to find c with $a < c < x$ such that

$$\frac{f(x)}{g(x)}=\frac{f(x)-f(a)}{g(x)-g(a)}=\frac{f'(c)}{g'(c)}$$

Then

$$\lim_{x\to a+}\frac{f(x)}{g(x)}=\lim_{c\to a+}\frac{f'(c)}{g'(c)}=\lim_{x\to a+}\frac{f'(x)}{g'(x)},$$

since we know the actual limit (not just the one sided limit) exists. Now repeat with $x < a$ to get the result.

Example: Evaluate $\lim_{x\to 0}\frac{1-\cos x}{x^2}=\frac{1}{2}$.

Solution: We have

$$\lim_{x\to 0}\frac{1-\cos x}{x^2}=\lim_{x\to 0}\frac{\sin x}{2x}=\frac{1}{2}$$

where the use of l'Hopital is justified since the second limit exists. Note that you can't differentiate top and bottom again, and still expect to get the correct answer; one of the hypotheses of l'Hopital is that the first quotient is of the 0/0 form.

Example: Use l'Hopital to establish the following:

$$\lim_{x\to 0}\frac{\sqrt{1+x}-1-x/2}{x^2}=\frac{1}{8}$$

Solution: We have

$$\lim_{x\to 0}\frac{\sqrt{1+x}-1-x/2}{x^2}$$

$$=\lim_{x\to 0}\frac{(1/2)(1+x)^{-1/2}-1/2}{2x}$$

$$=\lim_{x\to 0}\frac{(1/2)(-1/2)(1+x)^{-3/2}}{2}=\frac{-1}{8}$$

The use of l'Hopital is justified the second time, since the third limit exists; since we now know the second limit exists, the use of l'Hopital is justified the first time.

Infinite Limits

We can use Proposition to get results about infinite limits.

Example: Evaluate $\lim_{x\to\infty} x\log\left(1+\frac{1}{x}\right)$

Solution:

$$\lim_{x\to\infty} x\log\left(1+\frac{1}{x}\right)=\lim_{x\to\infty}\frac{\log(1+y)}{y}\text{ writing } y=1/x,$$

$$=\lim_{y\to 0}\frac{\log(1+y)}{y}=\lim_{y\to 0}\frac{\log(1+y)}{y}=1$$

The last step is valid, since the final limit exists by l'Hopital; note also that this gives another way of finding $a_n = (1 + 1/n)^n$.

l'Hopital's Rule: Infinite Limits

Proposition: Let f and g be functions such that

$$\lim_{x\to\infty} f(x) = \lim_{x\to\infty} g(x) = \infty,$$

and suppose that $\lim_{x\to\infty} \frac{f'(x)}{g'(x)}$ exists. Then

$$\lim_{x\to\infty} \frac{f(x)}{g(x)} = \lim_{x\to\infty} \frac{f'(x)}{g'(x)}$$

Proof: Pick $\in > 0$ and choose a such that

$$\lim_{x\to\infty} \left|\frac{f'(x)}{g'(x)}\right| - l < \in \quad \text{for all } x > a.$$

Then pick K such that if $x > K$, then $g(x) - g(a) \neq 0$. By Cauchy,

$$\frac{f'(c)}{g'(c)} = \frac{f(x)-f(a)}{g(x)-g(a)} \quad \text{for all } x > K.$$

Note that although c depends on x, we always have $c > a$. Then

$$\frac{f(x)}{g(x)} = \frac{f(x)-f(a)}{g(x)-g(a)} \cdot \frac{f(x)}{f(x)-f(a)} \cdot \frac{g(x)-g(a)}{g(x)}.$$

Chapter 3

Geometric Probability

Problems of the following type, and their solution techniques, were first studied in the 19th century, and the general topic became known as geometric probability.

- (Buffon's needle) What is the chance that a needle dropped randomly onto a floor marked with equally spaced parallel lines will cross one of the lines?
- What is the mean length of a random chord of a unit circle? (cf. Bertrand's paradox).
- What is the chance that three random points in the plane form an acute (rather than obtuse) triangle?
- What is the mean area of the polygonal regions formed when randomly-oriented lines are spread over the plane?

Since the late 20th century the topic has split into two topics with different emphases. Integral geometry sprang from the principle that the mathematically natural probability models are those that are invariant under certain transformation groups.

This topic emphasises systematic development of formulas for calculating expected values associated with the geometric objects derived from random points, and can in part be viewed as a sophisticated branch of multivariate calculus. Stochastic geometry emphasises the random geometrical objects themselves.

For instance: different models for random lines or for random tessalations of the plane; random sets formed by making points of a spatial Poisson process be (say) centres of discs.

Big O in Probability Notation

The order in probability notation is used in probability theory and statistical theory in direct parallel to the big-O notation which is standard in mathematics. Where the big-O notation deals with the convergence of sequences or sets of ordinary numbers, the order in probability notation deals with convergence of sets of random variables, where convergence is in the sense of convergence in probability.

For a set of random variables X_n and a corresponding set of constants a_n (both indexed by n which need not be discrete), the notation

$$X_n = o_p(a_n)$$

means that the set of values X_n/a_n converges to zero in probability as n approaches an appropriate limit. Equivalently, $X_n = o_p(a_n)$ can be written as $a_n\, o_p(1)$ where $X_n = o_p(1)$ is defined as,

$$\lim_{n\to\infty} P(|X_n| \geq \varepsilon) = 0,$$

for every positive ε.

The notation

$$X_n = O_p(a_n)$$

means that the set of values X_n/a_n is bounded in the limit in probability.

Example

If $\{X_n\}$ is a stochastic sequence such that each element has finite variance, then

$$X_n - E(X_n) = O_p(\sqrt{\operatorname{var}(X_n)})$$

(Theorem 14.4-1 in Bishop et al.)

Characteristic Function (Probability Theory)

In probability theory and statistics, the characteristic function of any real-valued random variable completely defines its probability distribution. If a random variable admits a probability density function, then the characteristic function is the Fourier transform of the probability density function. Thus it provides the basis of an alternative route to analytical results compared with working directly with probability density functions or cumulative distribution functions. There are particularly simple results for the characteristic functions of distributions defined by the weighted sums of random variables.

In addition to univariate distributions, characteristic functions can be defined for vector- or matrix-valued random variables, and can even be extended to more generic cases.

The characteristic function always exists when treated as a function of a real-valued argument, unlike the moment-generating function. There are relations between the behaviour of the characteristic function of a distribution and properties of the distribution, such as the existence of moments and the existence of a density function.

Introduction

The characteristic function provides an alternative way for describing a random variable. Similarly to the cumulative distribution function

$$F_X(x) = \mathrm{E}\left[1_{\{X \le x\}}\right],$$

where $1_{\{X \le x\}}$ is the indicator function — it is equal to 1 when $X \le x$, and zero otherwise. Which completely determines behaviour and properties of the probability distribution of the random variable X, the characteristic function

$$\varphi_X(t) = \mathrm{E}\left[e^{itX}\right]$$

also completely determines behaviour and properties of the probability distribution of the random variable X. The two approaches are equivalent in the sense that knowing one of the functions it is always possible to find the other, yet they both provide different insight for understanding the features of the random variable. However, in particular cases, there can be differences in whether these functions can be represented as expressions involving simple standard functions.

If a random variable admits a density function, then the characteristic function is its dual, in the sense that each of them is a Fourier transform of the other. If a random variable has a moment-generating function, then the domain of the characteristic function can be extended to the complex plane, and

$$\varphi_X(-it) = M_X(t).$$

Note however that the characteristic function of a distribution always exists, even when the probability density function or moment-generating function do not.

The characteristic function approach is particularly useful in analysis of linear combinations of independent random variables: a classical proof of the Central Limit Theorem uses characteristic functions and Lévy's continuity theorem. Another important application is to the theory of the decomposability of random variables.

Definition

For a scalar random variable X the characteristic function is defined as the expected value of e^{itX}, where i is the imaginary unit, and t – R is the argument of the characteristic function:

$$\begin{cases} \varphi_X : \mathrm{R} \to \mathrm{C} \\ \varphi_X(t) = \mathrm{E}\left[e^{itX}\right] = \int_{\mathrm{R}} e^{itx}\, dF_X(x) = \int_{\mathrm{R}} e^{itx} f_X(x) dx \end{cases}$$

Here F_X is the cumulative distribution function of X, and the integral is of the Riemann–Stieltjes kind. If random variable X has a probability density function f_X, then the characteristic function is its Fourier transform, and the last formula in parentheses is valid.

It should be noted though, that this convention for the constants appearing in the definition of the characteristic function differs from the usual convention for the Fourier transform. For example some authors define $\varphi_X(t) = \mathrm{E}{\geq}^2\pi^{itX}$, which is essentially a change of parameter. Other notation may be encountered in the literature: $\hat{p}$ as the characteristic function for a probability measure p, or $\hat{f}$ as the characteristic function corresponding to a density f.

Properties

- The characteristic function of a real-valued random variable always exists, since it is an integral of a bounded continuous function over a space whose measure is finite.
- A characteristic function is uniformly continuous on the entire space

- It is non-vanishing in a region around zero: $\varphi(0) = 1$.
- It is bounded: $|\varphi(t)| \leq 1$.
- It is Hermitian: $\varphi(-t) = \varphi(t)$. In particular, the characteristic function of a symmetric (around the origin) random variable is real-valued and even.
- There is a bijection between distribution functions and characteristic functions. That is, for any two random variables X_1, X_2

$$F_{X_1} = F_{X_2} \Leftrightarrow \varphi_{X_1} = \varphi_{X_2}$$

- If a random variable X has moments up to k-th order, then the characteristic function φ_X is k times continuously differentiable on the entire real line. In this case

$$\mathrm{E}[X^k] = (-i)^k \varphi_X^{(k)}(0).$$

- If a characteristic function φ_X has a k-th derivative at zero, then the random variable X has all moments up to k if k is even, but only up to $k - 1$ if k is odd.

$$\varphi_X^{(k)}(0) = i^k\, \mathrm{E}[X^k]$$

- If $X_1, \ldots, X_n$ are independent random variables, and $a_1, \ldots, a_n$ are some constants, then the characteristic function of the linear combination of X_i's is

$$\varphi_{a_1X_1+\ldots+a_nX_n}(t) = \varphi_{X_1}(a_1t) \cdot\ldots\cdot \varphi_{X_n}(a_nt).$$

One specific case would be the sum of two independent random variables X_1 and X_2 in which case one would have $\varphi_{X_1+X_2}(t) = \varphi_{X_1}(t) \cdot \varphi_{X_2}(t)$.

- The tail behaviour of the characteristic function determines the smoothness of the corresponding density function.

Continuity

The bijection stated above between probability distributions and characteristic functions is *continuous*. That is, whenever a sequence of distribution functions $F_j(x)$ converges (weakly) to some distribution $F(x)$, the corresponding sequence of characteristic functions $\varphi_j(t)$ will also converge, and the limit $\varphi(t)$ will correspond to the characteristic function of law F. More formally, this is stated as

Lévy's continuity theorem: A sequence X_j of n-variate random variables converges in distribution to random variable X if and only if the sequence φ_{X_j} converges pointwise to a function φ which is continuous at the origin. Then φ is the characteristic function of X.

This theorem is frequently used to prove the law of large numbers, and the central limit theorem.

Inversion Formulas

Since there is a one-to-one correspondence between cumulative distribution functions and characteristic functions, it is always possible to find one of these functions if we know the other one. The formula in definition of characteristic function allows us to compute φ when we know the distribution function F (or density f). If, on the other hand, we know the characteristic function φ and want to find the corresponding distribution function, then one of the following inversion theorems can be used.

Theorem. If characteristic function φ_X is integrable, then F_X is absolutely continuous, and therefore X has the probability density function given by

$$f_X(x) = F_{X'}(x) = \frac{1}{2\pi}\int_{\mathrm{R}} e^{-itx}\varphi_X(t)dt, \text{when } X \text{ is scalar;}$$

in multivariate case the pdf is understood as the Radon–Nikodym derivative of the distribution μ_X with respect to the Lebesgue measure λ:

$$f_X(x) = \frac{d\mu_X}{d\lambda}(x) = \frac{1}{(2\pi)^n}\int_{\mathrm{R}^n} e^{-i(t\cdot x)}\varphi_X(t)\lambda(dt).$$

Theorem (Lévy). If φ_X is characteristic function of distribution function F_X, two points $a<b$ are such that $\{x \mid a < x < b\}$ is a continuity set of μ_X (in the univariate case this condition is equivalent to continuity of F_X at points a and b), then

- If X is scalar:

$$F_X(b) - F_X(a) = \frac{1}{2\pi}\lim_{T\to\infty}\int_{-T}^{+T}\frac{e^{-ita} - e^{-itb}}{it}\varphi_X(t)dt,$$

- If X is a vector random variable:

$$\mu_X\left(\{a < x < b\}\right) = \frac{1}{(2\pi)^n} \lim_{T_1\to\infty} \cdots \lim_{T_n\to\infty}$$

$$\int_{-T_1\le t_1\le T_1} \cdots \int_{-T_n\le t_n\le T_n} \prod_{k=1}^{n} \left(\frac{e^{-it_k a_k} - e^{-it_k b_k}}{it_k} \right) \varphi_X(t)\lambda(dt_1 \times \cdots \times dt_n)$$

Theorem. If a is (possibly) an atom of X (in the univariate case this means a point of discontinuity of F_X) then

- If X is scalar:

$$F_X(a) - F_X(a-0) = \lim_{T\to\infty} \frac{1}{2T} \int_{-T}^{+T} e^{-ita} \varphi_X(t)dt$$

- If X is a vector random variable:

$$\mu_X(\{a\}) = \lim_{T_1\to\infty} \cdots \lim_{T_n\to\infty} \left(\prod_{k=1}^{n} \frac{1}{2T_k} \right) \int_{-T}^{T} e^{-i(t\cdot a)} \varphi_X(t)\lambda(dt)$$

Theorem (Gil-Pelaez). For a univariate random variable X, if x is a continuity point of F_X then

$$F_X(x) = \frac{1}{2} - \frac{1}{\pi} \int_0^\infty \frac{\operatorname{Im}[e^{-itx} \varphi_X(t)]}{t} dt.$$

The integral may be not Lebesgue-integrable; for example, when X is the discrete random variable that is always 0, it becomes the Dirichlet integral. Inversion formulas for multivariate distributions are available.

Criteria for Characteristic Functions

First note that the set of all characteristic functions is closed under certain operations:

- A convex linear combination $\sum_n a_n \varphi_n(t)$ (with $a_n \ge 0, \sum_n a_n = 1$) of a finite or a countable number of characteristic functions is also a characteristic function.
- The product of a finite number of characteristic functions is also a characteristic function. The same holds for an infinite product provided that it converges to a function continuous at the origin.

- If φ is a characteristic function and α is a real number, then $\overline{\varphi}$, Re(φ), $|\varphi|^2$, and $\varphi(\alpha t)$ are also characteristic functions.

It is well known that any non-decreasing càdlàg function F with limits $F(-\infty)=0, F(+\infty)=1$ corresponds to a cumulative distribution function of some random variable.

There is also interest in finding similar simple criteria for when a given function φ could be the characteristic function of some random variable.

The central result here is Bochner's theorem, although its usefulness is limited because the main condition of the theorem, non-negative definiteness, is very hard to verify. Other theorems also exist, such as Khinchine's, Mathias's, or Cramér's, although their application is just as difficult.

Pólya's theorem, on the other hand, provides a very simple convexity condition which is sufficient but not necessary. Characteristic functions which satisfy this condition are called Pólya-type. Bochner's theorem. An arbitrary function $\varphi : \mathrm{R}^n \to \mathrm{C}$ is the characteristic function of some random variable if and only if φ is positive definite, continuous at the origin, and if $\varphi(0) = 1$. Khinchine's criterion.

A complex-valued, absolutely continuous function φ, with $\varphi(0) = 1$, is a characteristic function if and only if it admits the representation

$$\varphi(t) = \int_{\mathrm{R}} g(t+\theta)\overline{g(\theta)}d\theta.$$

Mathias' theorem: A real-valued, even, continuous, absolutely integrable function φ, with $\varphi(0) = 1$, is a characteristic function if and only if

$$(-1)^n \left(\int_{\mathrm{R}} \varphi(pt) e^{-\frac{t^2}{2}} H_{2n}(t)dt \right) \geq 0$$

for $n = 0,1,2,\ldots$, and all $p > 0$. Here H_{2n} denotes the Hermite polynomial of degree $2n$.

Pólya's theorem. If φ is a real-valued, even, continuous function which satisfies the conditions

- $\varphi(0) = 1$,

- φ is convex for $t > 0$,
- $\varphi(-) = 0$,

then $\varphi(t)$ is the characteristic function of an absolutely continuous symmetric distribution.

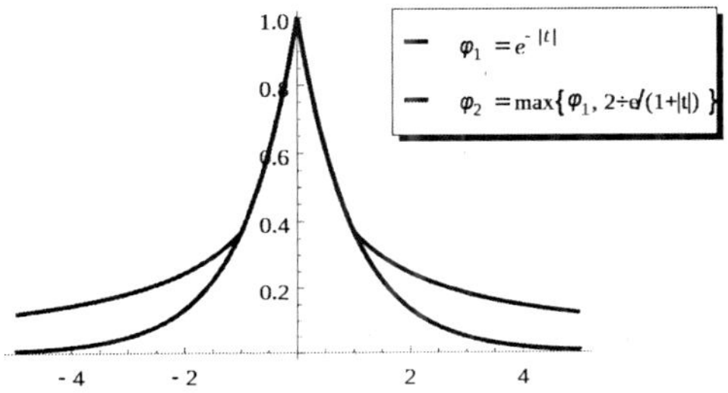

***Figure:** Pólya's theorem can be used to construct an example of two random variables whose characteristic functions coincide over a finite interval but are different elsewhere.*

Uses

Because of the continuity theorem, characteristic functions are used in the most frequently seen proof of the central limit theorem. The main trick involved in making calculations with a characteristic function is recognising the function as the characteristic function of a particular distribution.

Basic Manipulations of Distributions

Characteristic functions are particularly useful for dealing with linear functions of independent random variables. For example, if $X_1, X_2, \ldots, X_n$ is a sequence of independent (and not necessarily identically distributed) random variables, and

$$S_n = \sum_{i=1}^{n} a_i X_i,$$

where the a_i are constants, then the characteristic function for S_n is given by

$$\varphi_{S_n}(t) = \varphi_{X_1}(a_1 t)\varphi_{X_2}(a_2 t)\cdots\varphi_{X_n}(a_n t)$$

In particular, $\varphi_{X+Y}(t) = \varphi_X(t)\varphi_Y(t)$. To see this, write out the definition of characteristic function:

$$\varphi_{X+Y}(t) = \mathrm{E}\left[e^{it(X+Y)}\right] = \mathrm{E}\left[e^{itX}e^{itY}\right] = \mathrm{E}\left[e^{itX}\right]E\left[e^{itY}\right] = \varphi_X(t)\varphi_Y(t)$$

Observe that the independence of X and Y is required to establish the equality of the third and fourth expressions. Another special case of interest is when $a_i = 1/n$ and then S_n is the sample mean. In this case, writing X for the mean,

$$\varphi_{\bar{X}}(t) = \varphi_X\left(\tfrac{t}{n}\right)^n$$

Moments

Characteristic functions can also be used to find moments of a random variable. Provided that the nth moment exists, characteristic function can be differentiated n times and

$$\mathrm{E}\left[X^n\right] = i^{-n}\varphi_X^{(n)}(0) = i^{-n}\left[\frac{d^n}{dt^n}\varphi_X(t)\right]_{t=0}$$

For example, suppose X has a standard Cauchy distribution. Then $\varphi_X(t) = \geq^{|t|}$. See how this is not differentiable at $t = 0$, showing that the Cauchy distribution has no expectation.

Also see that the characteristic function of the sample mean X of n independent observations has characteristic function $\varphi_X(t) = (\geq^{|t|/n})^n = \geq^{|t|}$, using the result from the previous section. This is the characteristic function of the standard Cauchy distribution: thus, the sample mean has the same distribution as the population itself.

The logarithm of a characteristic function is a cumulant generating function, which is useful for finding cumulants; note that some instead define the cumulant generating function as the logarithm of the moment-generating function, and call the logarithm of the characteristic function the *second* cumulant generating function.

Data Analysis

Characteristic functions can be used as part of procedures for fitting probability distributions to samples of data. Cases where this provides a practicable option compared to other possibilities include fitting the stable distribution since closed form expressions for the density are not available which makes implementation of maximum likelihood estimation difficult.

Estimation procedures are available which match the theoretical characteristic function to the empirical characteristic function, calculated from the data.

Paulson et al. (1975) and Heathcote (1977) provide some theoretical background for such an estimation procedure. In addition, Yu (2004) describes applications of empirical characteristic functions to fit time series models where likelihood procedures are impractical.

Example

The Gamma distribution with scale parameter θ and a shape parameter k has the characteristic function

$$(1-\theta it)^{-k}.$$

Now suppose that we have

$$X \sim \Gamma(k_1,\theta) \text{ and } Y \sim \Gamma(k_2,\theta)$$

with X and Y independent from each other, and we wish to know what the distribution of $X + Y$ is. The characteristic functions are

$$\varphi_X(t) = (1-\theta it)^{-k_1}, \qquad \varphi_Y(t) = (1-\theta it)^{-k_2}$$

which by independence and the basic properties of characteristic function leads to

$$\varphi_{X+Y}(t) = \varphi_X(t)\varphi_Y(t) = (1-\theta it)^{-k_1}(1-\theta it)^{-k_2} = \left(1-\theta it\right)^{-(k_1+k_2)}.$$

This is the characteristic function of the gamma distribution scale parameter θ and shape parameter $k_1 + k_2$, and we therefore conclude

$$X + Y \sim \Gamma(k_1 + k_2,\theta)$$

The result can be expanded to n independent gamma distributed random variables with the same scale parameter and we get

$$\forall i \in \{1,\ldots,n\} : X_i \sim \Gamma(k_i,\theta) \qquad \Rightarrow \qquad \sum_{i=1}^{n} X_i \sim \Gamma\left(\sum_{i=1}^{n} k_i,\theta\right).$$

Entire Characteristic Functions

As defined above, the argument of the characteristic function is treated as a real number: however, certain aspects of the theory of characteristic functions are advanced by extending the definition into the complex plane by analytical continuation, in cases where this is possible.

Related Concepts

Related concepts include the moment-generating function and the probability-generating function. The characteristic function exists for all probability distributions. However this is not the case for moment generating function.

The characteristic function is closely related to the Fourier transform: the characteristic function of a probability density function $p(x)$ is the complex conjugate of the continuous Fourier transform of $p(x)$.

$$\varphi_X(t) = \langle e^{itX} \rangle = \int_{\mathrm{R}} e^{itx} p(x)dx = \overline{\left(\int_{\mathrm{R}} e^{-itx} p(x)dx\right)} = \overline{P(t)},$$

where $P(t)$ denotes the continuous Fourier transform of the probability density function $p(x)$. Likewise, $p(x)$ may be recovered from $\varphi_X(t)$ through the inverse Fourier transform:

$$p(x) = \frac{1}{2\pi}\int_{\mathrm{R}} e^{itx} P(t)dt = \frac{1}{2\pi}\int_{\mathrm{R}} e^{itx} \overline{\varphi_X(t)}dt.$$

Indeed, even when the random variable does not have a density, the characteristic function may be seen as the Fourier transform of the measure corresponding to the random variable.

Chebyshev's Inequality

In probability theory, Chebyshev's inequality guarantees that in any probability distribution, "nearly all" values are close to the mean — the precise statement being that no more than $1/k^2$ of the distribution's values can be more than k standard deviations away from the mean.

The inequality has great utility because it can be applied to completely arbitrary distributions (unknown except for mean and variance), for example it can be used to prove the weak law of large numbers. The term *Chebyshev's inequality* may also refer to the Markov's inequality, especially in the context of analysis.

History

The theorem is named after Russian mathematician Pafnuty Chebyshev, although it was first formulated by his friend and colleague Irénée-Jules Bienaymé. The theorem was first stated without proof by Bienaymé in 1853 and later also without proof by Chebyshev in 1874. His student Andrey Markov provided a proof in his 1884 PhD thesis.

Statement

Chebyshev's inequality is usually stated for random variables, but can be generalised to a statement about measure spaces.

Measure-theoretic Statement

Let (X, Σ, μ) be a measure space, and let f be an extended real-valued measurable function defined on X. Then for any real number $t > 0$,

$$\mu(\{x \in X : | f(x) | \geq t\}) \leq \frac{1}{t^2} \int_X |f|^2 \, d\mu.$$

More generally, if g is an extended real-valued measurable function, nonnegative and nondecreasing on the range of f, then

$$\mu(\{x \in X : f(x) \geq t\}) \leq \frac{1}{g(t)} \int_X g \circ f \, d\mu.$$

The previous statement then follows by defining $g(t)$ as t^2 if $t \geq 0$ and *0* otherwise, and taking $| f |$ instead of f.

Example

Suppose we randomly select a journal article from a source with an average of 1000 words per article, with a standard deviation of 200 words.

We can then infer that the probability that it has between 600 and 1400 words (i.e. within $k = 2$ SDs of the mean) must be more than 75%, because there is less than $^1/_{k2}$

= 14 chance to be outside that range, by Chebyshev's inequality. But if we additionally know that the distribution is normal, we can say that is a 75% chance the word count is between 770 and 1230 (which is an even tighter bound).

Note: This example should be treated with caution as the inequality is only stated for probability distributions rather than for finite sample sizes. The inequality has since been extended to apply to finite samp.

Sharpness of Bounds

As shown in the example above, the theorem will typically provide rather loose bounds. However, the bounds provided by Chebyshev's inequality cannot, in general (remaining sound for variables of arbitrary distribution), be improved upon. For example, for any $k \geq 1$, the following example meets the bounds exactly.

$$X = \begin{cases} -1, & \text{with probability } \dfrac{1}{2k^2} \\ 0, & \text{with probability } 1 - \dfrac{1}{k^2} \\ 1, & \text{with probability } \dfrac{1}{2k^2} \end{cases}$$

For this distribution, mean $\mu = 0$ and standard deviation $\sigma = 1k$, so

$$\Pr(|X - \mu| \geq k\sigma) = \Pr(|X| \geq 1) = \frac{1}{k^2}.$$

Equality holds only for distributions that are a linear transformation of this one.

Proof (of the two-sided Version)

Probabilistic proof

Markov's inequality states that for any real-valued random variable Y and any positive number a, we have $\Pr(|Y| > a) \leq \mathrm{E}(|Y|)/a$. One way to prove Chebyshev's inequality is to apply Markov's inequality to the random variable $Y = (X - \mu)^2$ with $a = (\sigma k)^2$. It can also be proved directly. For any event A, let I_A be the indicator random variable of A, i.e. I_A equals 1 if A occurs and 0 otherwise. Then

$$\Pr(|X - \mu| \geq k\sigma) = \mathrm{E}(I_{|X-\mu| \geq k\sigma}) = \mathrm{E}(I_{[(X-\mu)/(k\sigma)]^2 \geq 1})$$

$$[6pt] \quad \leq \mathrm{E}\left(\left(\frac{X-\mu}{k\sigma}\right)^2\right) = \frac{1}{k^2}\frac{\mathrm{E}((X-\mu)^2)}{\sigma^2} = \frac{1}{k^2}.$$

The direct proof shows why the bounds are quite loose in typical cases: the number 1 to the left of –≥– is replaced by [(X –

$\mu)/(k\sigma)]^2$ to the right of $-\geq-$ whenever the latter exceeds 1. In some cases it exceeds 1 by a very wide margin.

Measure-theoretic Proof

Fix t and let A_t be defined as $A_t = \{x \in X\, f(x) \geq t\}$ and let 1_{A_t} be the indicator function of the set . Then, it is easy to check that, for any,

$$0 \leq g(t)1_{A_t} \leq g(f(x))1_{A_t},$$

since g is nondecreasing on the range of f, and therefore,

$$\begin{aligned} g(t)\mu(A_t) &= \int_X g(t)1_{A_t}\, d\mu \\ &\leq \int_{A_t} g^\circ f\, d\mu \\ &\leq \int_X g^\circ f\, d\mu. \end{aligned}$$

The desired inequality follows from dividing the above ineq-uality by $g(t)$.

Extensions

Several extensions of Chebyshev's inequality have been developed.

Asymmetric two-sided Case

An asymmetric two-sided version of this inequality is also known. When the distribution is known to be symmetric

$$P(k_1 < X < k_2) \geq 1 - \frac{4\sigma^2}{(k_2 - k_1)^2}$$

where σ^2 is the variance.

When the distribution is asymmetric or is unknown

$$P(k_1 < X < k_2) \geq \frac{4[(\mu - k_1)(k_2 - \mu) - \sigma^2]}{(k_2 - k_1)^2},$$

where σ^2 is the variance and μ is the mean.

Bivariate Case

A version for the bivariate case is known. Let X_1 and X_2 be two random variables with means and finite variances of μ_1 and μ_2 and σ_1 and σ_2 respectively. Then

$$P(k_{11} \le X_1 \le k_{12}, k_{21} \le X_2 \le k_{22}) \ge 1 - \sum T_i$$

where for i = 1,2,

$$T_i = \frac{4\sigma_i^2 + [2\mu_i - (k_{i1} + k_{i2})]^2}{(k_{i2} - k_{i1})}$$

Two Correlated Variables

Berge derived an inequality for two correlated variables X_1 and X_2. Let ρ be the correlation coefficient between X_1 and X_2 and let σ_i^2 be the variance of X_i. Then

$$P\left(\bigcap_{i=1}^{2}\left[\frac{|X_i - \mu_i|}{\sigma_i} < k\right]\right) \ge 1 - \frac{1+\sqrt{1-\rho^2}}{k^2}$$

Lal later obtained an alternative bound

$$P\left(\bigcap_{i=1}^{2}\left[\frac{|X_i - \mu_i|}{\sigma_i} \le k_i\right]\right) \ge 1 - \frac{k_1^2 + k_2^2 + \sqrt{(k_1^2 + k_2^2)^2 - 4k_1^2 k_2^2 \rho}}{2(k_1 k_2)^2}$$

Isii derived a further generalisation. Let

$$Z = P\left(\bigcap_{i=1}^{2}(-k_1 < X_i < k_2)\right)$$

with $0 < k_1 \le k_2$.

There are now three cases.

Case A: If $2k_1^2 > 1 - \rho$ and $k_2 - k_1 \ge 2\lambda$ where

$$\lambda = \frac{k_1(1+\rho) + \sqrt{(1-\rho^2)(k_1^2 + \rho)}}{2k_1 - 1 + \rho}$$

then

$$Z \le \frac{2\lambda^2}{2\lambda^2 + 1 + \rho}$$

Case B: If the conditions in case A are not met but $k_1 k_2 \ge 1$ and

$$2(k_1 k_2 - 1)^2 \ge 2(1-\rho^2) + (1-\rho)(k_2 - k_1)^2$$

then

$$Z \le \frac{(k_2 - k_1)^2 + 4 + \sqrt{16(1-\rho^2) + 8(1-\rho)(k_2 - k_1)}}{(k_1 + k_2)^2}$$

Case C: If the conditions in cases A or B are not met there is no universal bound other than 1.

Multivariate Case

The general case is known as the Birnbaum–Raymond–Zuckerman inequality after the authors who proved it for two dimensions.

$$P\left[\sum_{i=1}^{n}\frac{(X_i-\mu_i)^2}{\sigma_i^2 t_i^2}\geq k^2\right]\leq\frac{1}{k^2}\sum_{i=1}^{n}\frac{1}{t_i^2}$$

where X_i is the ith random variable, μ_i is the ith mean and σ_i^2 is the ith variance.

If the variables are independent this inequality can be shar-pened.

$$P\left[\bigcap_{i=1}^{n}\frac{|X_i-\mu_i|}{\sigma_i}\leq k_i\right]\geq\prod(1-\frac{1}{k_i^2})$$

Olkin and Pratt derived an inequality for n correlated variables.

$$P\left(\bigcap_{i=1}^{n}\frac{|X_i-\mu_i|}{\sigma_i}<k_i\right)\geq 1-\frac{[\sqrt{u}+\sqrt{n-1}\sqrt{n\sum\frac{1}{k_i^2}-u}]^2}{n^2}$$

where the sum is taken over the n variables and

$$u=\sum_{i=1}^{n}\frac{1}{k_i^2}+2\sum_{i=1}^{n}\sum_{i<j}\frac{\rho_{ij}}{k_i k_j}$$

where ρ_{ij} is the correlation between X_i and X_j

Olkin and Pratt's inequality was subsequently generalised by Godwin.

Finite Samples

Saw *et al* extended Chebyshev's inequality to cases where the population mean and variance are not known but are instead replaced by their sample estimates.

$$P(|X-m|\geq ks)\leq\frac{g_{N+1}\left(\frac{Nk^2}{N-1+k^2}\right)}{N+1}\left(\frac{N}{N+1}\right)^{1/2}$$

where N is the sample size, m is the sample mean, k is a constant and s is the sample standard deviation. g(x) is defined as follows:

Let $x \geq 1$, $Q = N + 1$, and R be the greatest integer less than Q / x. Let

$$a^2 = \frac{Q(Q-R)}{1+R(Q-R)}$$

Now

$$g_Q(x) = R \quad \text{if } R \text{ is even}$$

$$g_Q(x) = R \quad \text{if } R \text{ is odd and } x < a^2$$

$$g_Q(x) = R - 1 \quad \text{if } R \text{ is odd and } x \geq a^2$$

This inequality holds when the population moments do not exist and when the sample is weakly exchangeably distributed. Kaban gives a somewhat less complex version of this inequality.

$$P(|X-m| \geq ks) \leq \frac{1}{[N(N+1)]^{1/2}}\left[\left(\frac{N-1}{k^2}+1\right)\right]$$

If the standard deviation is a multiple of the mean then a further inequality can be derived,

$$P(|X-m| \geq ks) \leq \frac{N-1}{N}\frac{1}{k^2}\frac{s^2}{m^2}+\frac{1}{N}$$

A table of values for the Saw–Yang–Mo inequality for finite sample sizes ($n < 100$) has been determined by Konijn.

For fixed N and large m the Saw–Yang–Mo inequality is approximately

$$P(|X-m| \geq ks) \leq \frac{1}{N+1}$$

Beasley *et al* have suggested a modification of this inequality

$$P(|X-m| \geq ks) \leq \frac{1}{k^2(N+1)}$$

In empirical testing this modification is conservative but appears to have low statistical power. Its theoretical basis currently remains unexplored.

Dependence of Sample Size

The bounds these inequalities give on a finite sample are less tight than those the Chebyshev inequality gives for a distribution. To illustrate this let the sample size $n = 100$ and let $k = 3$. Chebyshev's inequality states that approximately 11.11% of the distribution will lie outside these limits. Kabán's version of the inequality for a finite sample states that approximately 12.05% of the sample lies outside these limits. The dependence of the confidence intervals on sample size is further illustrated below.

For $N = 10$, the 95% confidence interval is approximately ±13.5789 standard deviations.

For $N = 100$ the 95% confidence interval is approximately ±4.9595 standard deviations; the 99% confidence interval is approximately ±140.0 standard deviations.

For $N = 500$ the 95% confidence interval is approximately ±4.5574 standard deviations; the 99% confidence interval is approximately ±11.1620 standard deviations.

For $N = 1000$ the 95% and 99% confidence intervals are approximately ±4.5141 and approximately ±10.5330 standard deviations respectively.

The Chebyshev inequality for the distribution gives 95% and 99% confidence intervals of approximately ±4.472 standard deviations and ±10 standard deviations respectively.

Comparative Bounds

Although Chebyshev's inequality is the best possible bound for an arbitrary distribution, this is not necessarily true for finite samples. Samuelson's inequality states that all values of a sample will lie within $-(N - 1)$ standard deviations of the mean. Chebyshev's bound improves as the sample size increases.

When $N = 10$, Samuelson's inequality states that all members of the sample lie within 3 standard deviations of the mean: in contrast Chebyshev's states that 95% of the sample lies within 13.5789 standard deviations of the mean.

When $N = 100$, Samuelson's inequality states that all members of the sample lie within approximately 9.9499 standard deviations of the mean: Chebyshev's states that 99% of the sample lies within 140.0 standard deviations of the mean.

When $N = 500$, Samuelson's inequality states that all members of the sample lie within approximately 22.3383 standard deviations of the mean: Chebyshev's states that 99% of the sample lies within 11.1620 standard deviations of the mean. It is likely that better bounds for finite samples than these exist.

Sharpened Bounds

Chebyshev's inequality is important because of its applicability to any distribution. As a result of its generality it may not (and usually does not) provide as sharp a bound as alternative methods that can be used if the distribution of the random variable is known.

To improve the sharpness of the bounds provided by Chebyshev's inequality a number of methods have been developed.

Standardised Variables

Sharpened bounds can be derived by first standardising the random variable. Let X be a random variable with finite variance $Var(x)$. Let Z be the standardised form defined as

$$Z = \frac{X - \mathrm{E}(X)}{\mathrm{Var}(X)^{1/2}}$$

Cantelli's lemma is then

$$P(Z \geq k) \leq \frac{1}{1+k^2}$$

This inequality is sharp and is attained by k and $-1/k$ with probability $1/(1 + k^2)$ and $k^2/(1 + k^2)$ respectively. If $k > 1$ and the distribution of X is symmetric then we have

$$P(Z \geq k) \leq \frac{1}{2k^2}.$$

Equality holds if and only if $Z = -k$, 0 or k with probabilities $1 / 2 k^2$, $1 - 1 / k^2$ and $1 / 2 k^2$ respectively. An extension to a two-sided inequality is also possible.

Let $u, v > 0$. Then we have

$$P(Z \leq -u \text{ or } Z \geq v) \leq \frac{4+(u-v)^2}{(u+v)^2}.$$

Semivariances

An alternative method of obtaining sharper bounds is through the use of semivariances (partial moments). The upper (σ_+^2) and lower (σ_-^2) semivariances are defined

$$\sigma_+^2 = \frac{\sum (x-m)^2}{n-1}$$

$$\sigma^2 = \sigma_+^2 + \sigma_-^2$$

where m is the arithmetic mean of the sample, n is the number of elements in the sample and the sum for the upper (lower) semivariance is taken over the elements greater (less) than the mean. The variance of the sample is the sum of the two semivariances

$$\sigma^2 = \sigma_+^2 + \sigma_-^2$$

In terms of the lower semivariance Chebyshev's inequality can be written

$$\Pr(x \le m - a\sigma_-) \le \frac{1}{a^2}$$

Putting

$$a = \frac{k\sigma}{\sigma_-}$$

Chebyshev's inequality can now be written

$$\Pr(x \le m - k\sigma) \le \frac{1}{k^2}\frac{\sigma_-^2}{\sigma^2}$$

A similar result can also be derived for the upper semivariance.

If we put

$$\sigma_u^2 = \max(\sigma_-^2, \sigma_+^2),$$

Chebyshev's inequality can be written

$$\Pr(|\, x \le m - k\sigma \,|) \le \frac{1}{k^2}\frac{\sigma_u^2}{\sigma^2}.$$

Because $\sigma_u^2 \le \sigma^2$, use of the semivariance sharpens the original inequality.

If the distribution is known to be symmetric, then

$$\sigma_+^2 = \sigma_-^2 = \frac{1}{2}\sigma^2$$

and

$$\Pr(x \le m - k\sigma) \le \frac{1}{2k^2}.$$

This result agrees with that derived using standardised variables.

Note: The inequality with the lower semivariance has been found to be of use in estimating downside risk in finance and agriculture.

Cantelli's Inequality

Cantelli's inequality due to Francesco Paolo Cantelli states that for a real random variable (X) with mean (μ) and variance (σ^2)

$$P(X - \mu \ge a) \le \frac{\sigma^2}{\sigma^2 + a^2}$$

where $a \ge 0$.

This inequality can be used to prove a one tailed variant of Chebyshev's inequality with $k > 0$

$$\Pr(X - \mu \ge k\sigma) \le \frac{1}{1 + k^2}.$$

The bound on the one tailed variant is known to be sharp. To see this consider the random variable X that takes the values

$$X = 1 \text{with probability} \frac{\sigma^2}{1 + \sigma^2}$$

$$X = -\sigma^2 \text{with probability} \frac{1}{1 + \sigma^2}.$$

Then $E(X) = 0$ and $E(X^2) = \sigma^2$ and $P(X < 1) = 1 / (1 + \sigma^2)$.

An application – distance between the mean and the median

The one-sided variant can be used to prove the proposition that for probability distributions having an expected value and a median, the mean and the median can never differ from each other by more than one standard deviation. To express this in symbols let μ, ν, and σ be respectively the mean, the median, and the standard deviation. Then

$$|\mu - \nu| \leq \sigma.$$

There is no need to assume that the variance is finite because this inequality is trivially true if the variance is infinite. The proof is as follows. Setting $k = 1$ in the statement for the one-sided inequality gives:

$$\Pr(X - \mu \geq \sigma) \leq \frac{1}{2}.$$

Changing the sign of X and of μ, we get

$$\Pr(X \leq \mu - \sigma) \geq \frac{1}{2}.$$

Thus the median is within one standard deviation of the mean.

Chapter 4

Algebraic Geometry

Algebraic geometry is a branch of mathematics which combines techniques of abstract algebra, especially commutative algebra, with the language and the problems of geometry.

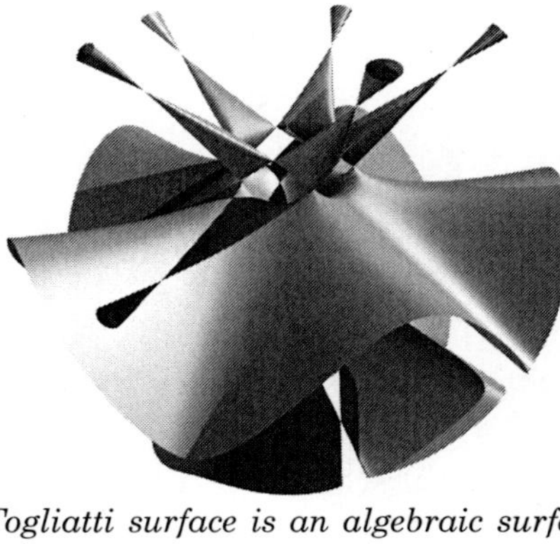

Figure: *This Togliatti surface is an algebraic surface of degree five*

It occupies a central place in modern mathematics and has multiple conceptual connections with such diverse fields as complex analysis, topology and number theory. Initially a study of systems of polynomial equations in several variables, the subject of algebraic geometry starts where equation solving leaves off, and it becomes even more important to understand the intrinsic properties of the totality of solutions of a system of equations, than to find a specific solution; this leads into some of the deepest areas in all of mathematics, both conceptually and in terms of technique.

The fundamental objects of study in algebraic geometry are algebraic varieties, geometric manifestations of solutions of systems of polynomial equations. Plane algebraic curves, which include lines, circles, parabolas, ellipses, hyperbolas, cubic curves like elliptic curves and quartic curves like lemniscates, and Cassini ovals, are some of the most studied classes of algebraic varieties. A point of the plane belongs to an algebraic curve if its coordinates satisfy a given polynomial equation. Basic questions involve the study of the points of special interest like the singular points, the inflexion points and the points at infinity. More advanced questions involve the topology of the curve and relations between the curves given by different equations. In the 20th century, algebraic geometry has split into several subareas.

- The main stream of algebraic geometry is devoted to the study of the complex points of the algebraic varieties and more generally to the points with coordinates in an algebraically closed field.
- The study of the points of an algebraic variety with coordinates in the field of the rational numbers or in a number field became algebraic number theory.
- The study of the real points of an algebraic variety is the subject of real algebraic geometry.
- A large part of singularity theory is devoted to the singularities of algebraic varieties.
- With the rise of the computers, a computational algebraic geometry area has emerged, which lies at the intersection of algebraic geometry and computer algebra. It consists essentially in developing algorithms and software for studying and finding the properties of explicitly given algebraic varieties.
- Much of the development of the main stream of algebraic geometry in the 20th century occurred within an abstract algebraic framework, with increasing emphasis being placed on "intrinsic" properties of algebraic varieties not dependent on any particular way of embedding the variety in an ambient coordinate space; this parallels developments in topology, differential and complex geometry. One key achievement of this abstract algebraic geometry is Grothendieck's scheme theory which allows one to use sheaf theory to study algebraic

varieties in a way which is very similar to its use in the study of differential and analytic manifolds. This is obtained by extending the notion of point: In classical algebraic geometry, a point of an affine variety may be identified, through Hilbert's Nullstellensatz, with a maximal ideal of the coordinate ring, while the points of the corresponding affine scheme are all prime ideals of this ring. This means that a point of such a scheme may be either a usual point or a subvariety. This approach also enables a unification of the language and the tools of classical algebraic geometry, mainly concerned with complex points, and of algebraic number theory. Wiles's proof of the longstanding conjecture called Fermat's last theorem is an example of the power of this approach.

Basis Notions

Zeros of Simultaneous Polynomials

In classical algebraic geometry, the main objects of interest are the vanishing sets of collections of polynomials, meaning the set of all points that simultaneously satisfy one or more polynomial equations. For instance, the two-dimensional sphere in three-dimensional Euclidean space R^3 could be defined as the set of all points (x,y,z) with

$$x^2 + y^2 + z^2 - 1 = 0.$$

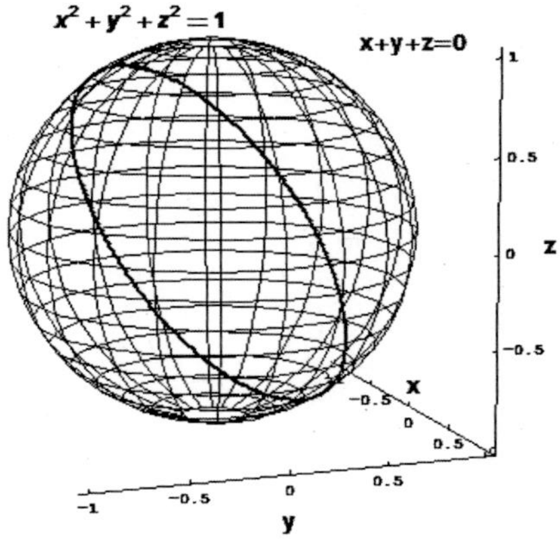

Figure: *Sphere and slanted circle*

A "slanted" circle in R^3 can be defined as the set of all points (x,y,z) which satisfy the two polynomial equations

$$x^2 + y^2 + z^2 - 1 = 0,$$

$$x + y + z = 0.$$

Affine Varieties

First we start with a field k. In classical algebraic geometry, this field was always the complex numbers C, but many of the same results are true if we assume only that k is algebraically closed. We consider the affine space of dimension n over k, denoted $A^n(k)$ (or more simply A^n, when k is clear from the context). When one fixes a coordinates system, one may identify $A^n(k)$ with k^n. The purpose of not working with k^n is to emphasise that one "forgets" the vector space structure that k^n carries.

A function $f : A^n \to A^1$ is said to be *polynomial* (or *regular*) if it can be written as a polynomial, that is, if there is a polynomial p in $k[x_1,...,x_n]$ such that $f(M) = p(t_1,...,t_n)$ for every point M with coordinates $(t_1,...,t_n)$ in A^n. The property of a function to be polynomial (or regular) does not depend on the choice of a coordinate system in A^n.

Regular functions on affine n-space are thus exactly the same as polynomials over k in n variables. We will refer to the set of all regular functions on A^n as $k[A^n]$.

We say that a polynomial *vanishes* at a point if evaluating it at that point gives zero. Let S be a set of polynomials in $k[A^n]$. The *vanishing set of S* (or *vanishing locus*) is the set $V(S)$ of all points in A^n where every polynomial in S vanishes. In other words,

$$V(S) = \{(t_1,\ldots,t_n) \mid \forall p \in S, p(t_1,\ldots,t_n) = 0\}.$$

A subset of A^n which is $V(S)$, for some S, is called an *algebraic set*. The V stands for *variety* (a specific type of algebraic set to be defined below).

Given a subset U of A^n, can one recover the set of polynomials which generate it? If U is *any* subset of A^n, define $I(U)$ to be the set of all polynomials whose vanishing set contains U. The I stands for ideal: if two polynomials f and g both vanish on U, then $f+g$ vanishes on U, and if h is any polynomial, then hf vanishes on U, so $I(U)$ is always an ideal of $k[A^n]$.

Two natural questions to ask are:

- Given a subset U of A^n, when is $U = V(I(U))$?
- Given a set S of polynomials, when is $S = I(V(S))$?

The answer to the first question is provided by introducing the Zariski topology, a topology on A^n whose closed sets are the algebraic sets, and which directly reflects the algebraic structure of $k[A^n]$. Then $U = V(I(U))$ if and only if U is an algebraic set or equivalently a Zariski-closed set. The answer to the second question is given by Hilbert's Nullstellensatz. In one of its forms, it says that $I(V(S))$ is the radical of the ideal generated by S. In more abstract language, there is a Galois connection, giving rise to two closure operators; they can be identified, and naturally play a basic role in the theory; the example is elaborated at Galois connection.

For various reasons we may not always want to work with the entire ideal corresponding to an algebraic set U. Hilbert's basis theorem implies that ideals in $k[A^n]$ are always finitely generated.

An algebraic set is called *irreducible* if it cannot be written as the union of two smaller algebraic sets. Any algebraic set is a finite union of irreducible algebraic sets and this decomposition is unique. Thus its elements are called the *irreducible components* of the algebraic set. An irreducible algebraic set is also called a *variety*. It turns out that an algebraic set is a variety if and only if it may be defined as the vanishing set of a prime ideal of the polynomial ring.

Some authors do not make a clear distinction between algebraic sets and varieties and use *irreducible variety* to make the distinction when needed.

Regular Functions

Just as continuous functions are the natural maps on topological spaces and smooth functions are the natural maps on differentiable manifolds, there is a natural class of functions on an algebraic set, called *regular functions* or *polynomial functions*. A regular function on an algebraic set V contained in A^n is the restriction to V of a regular function on A^n. For an algebraic set defined on the field of the complex numbers, the regular functions are smooth and even analytic.

It may seem unnaturally restrictive to require that a regular function always extend to the ambient space, but it is very similar to the situation in a normal topological space, where the Tietze extension theorem guarantees that a continuous function on a closed subset always extends to the ambient topological space.

Just as with the regular functions on affine space, the regular functions on V form a ring, which we denote by $k[V]$. This ring is called the *coordinate ring of V*. Since regular functions on V come from regular functions on A^n, there is a relationship between the coordinate rings. Specifically, if a regular function on V is the restriction of two functions f and g in $k[A^n]$, then $f - g$ is a polynomial function which is null on V and thus belongs to $I(V)$. Thus $k[V]$ may be identified with $k[A^n]/I(V)$.

Morphism of Affine Varieties

Using regular functions from an affine variety to A^1, we can define regular maps from one affine variety to another. First we will define a regular map from a variety into affine space: Let V be a variety contained in A^n. Choose m regular functions on V, and call them $f_1, ..., f_m$. We define a *regular map* f from V to A^m by letting $f = (f_1, ..., f_m)$. In other words, each f_i determines one coordinate of the range of f.

If V' is a variety contained in A^m, we say that f is a *regular map* from V to V' if the range of f is contained in V'.

The definition of the regular maps apply also to algebraic sets. The regular maps are also called *morphisms*, as they make the collection of all affine algebraic sets into a category, where the objects are the affine algebraic sets and the morphisms are the regular maps. The affine varieties is a subcategory of the category of the algebraic sets.

Given a regular map g from V to V' and a regular function f of $k[V']$, then f"g"$k[V]$. The map f'!f"g is a ring homomorphism from $k[V']$ to $k[V]$. Conversely, every ring homomorphism from $k[V']$ to $k[V]$ defines a regular map from V to V'.

This defines an equivalence of categories between the category of algebraic sets and the opposite category of the finitely generated reduced k-algebras. This equivalence is one of the starting points of scheme theory.

Rational Function and Birational Equivalence

Contrarily to the preceding ones, this section concerns only varieties and not algebraic sets. On the other hand the definitions extend naturally to projective varieties (next section), as an affine variety and its projective completion have the same field of functions.

If V is an affine variety, its coordinate ring is an integral domain and has thus a field of fractions which is denoted $k(V)$ and called the *field of the rational functions* on V or, shortly, the *function field* of V. Its elements are the restrictions to V of the rational functions over the affine space containing V. The domain of a rational function f is not V but the complement of the subvariety (a hypersurface) where the denominator of f vanishes. Like for regular maps, one may define a *rational map* from a variety V to a variety V'. Like for the regular maps, the rational maps from V to V' may be identified to the field homomorphisms from $k(V')$ to $k(V)$.

Two affine varieties are *birationally equivalent* if there two rational functions between them which are inverse one to the other in the regions where both are defined. Equivalently, they are birationally equivalent if their function fields are isomorphic. An affine variety is a *rational variety* if it is birationally equivalent to an affine space. This means that the variety admits a rational parameterisation. For example, the circle of equation $x^2 + y^2 - 1 = 0$ is a rational curve, as it has the parameterisation

$$x = \frac{2t}{1+t^2}$$

$$y = \frac{1-t^2}{1+t^2},$$

which may also be viewed as a rational map from the line to the circle.

The problem of resolution of singularities is to know if every algebraic variety is birationally equivalent to a variety whose projective completion is non singular. It has been positively solved in characteristic 0 by Hironaka in 1964 and is yet unsolved in finite characteristic.

Projective Variety

Many properties of the affine varieties depend on their behaviour "at infinity".

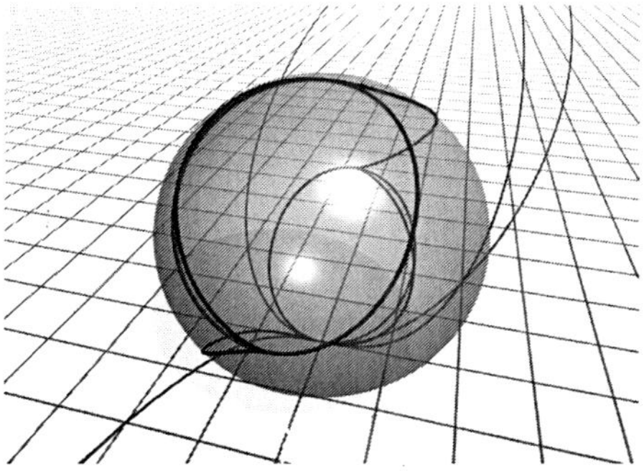

Figure: *parabola ($y = x^2$, red) and cubic ($y = x^3$, blue) in projective space*

For example, consider the variety $V(y - x^2)$. If we draw it, we get a parabola. As x increases, the slope of the line from the origin to the point (x, x^2) becomes larger and larger. As x decreases, the slope of the same line becomes smaller and smaller.

Compare this to the variety $V(y - x^3)$. This is a cubic curve. As x increases, the slope of the line from the origin to the point (x, x^3) becomes larger and larger just as before. But unlike before, as x decreases, the slope of the same line again becomes larger and larger. So the behaviour "at infinity" of $V(y - x^3)$ is different from the behaviour "at infinity" of $V(y - x^2)$.

The consideration of the *projective completion* of the two curves, which is their prolongation "at infinity" in the projective plane, allows to quantify this difference: the point at infinity of the parabola is a regular point, whose tangent is the line at infinity, while the point at infinity of the cubic curve is a cusp. Also, both curves are rational, as they are parameterised by x, and Riemann-Roch theorem implies that the cubic curve must have a singularity, which must be at infinity, as all its points in the affine space are regular.

Thus many of the properties of the algebraic varieties, including birational equivalence and all the topological properties depends on the behaviour "at infinity" and, thus imply to study the varieties in the projective space. Furthermore, the introduction of projective techniques made many theorems in algebraic geometry simpler and sharper: For example, Bézout's theorem on the number of intersection points between two varieties can be stated in its sharpest form only in projective space. For these reasons, projective space plays a fundamental role in algebraic geometry.

Nowadays, the *projective space* P^n of dimension n is usually defined as the set of the lines passing through a point, considered as the origin, in the affine space of dimension n+1, or equivalently to the set of the vector lines in a vector space of dimension n+1. When a coordinate system has been chosen in the space of dimension n+1, all the points of a line have the same set of coordinates, up to the multiplication by an element of k. This defines the homogeneous coordinates of a point of P^n as a sequence of n+1 elements of the base field k, defined up to the multiplication by a non zero element of k (the same for the whole sequence).

Given a polynomial in n+1 variables, it vanishes at all the point of a line passing through the origin if and only if it is homogeneous. In this case, one says that the polynomial *vanishes* at the corresponding point of P^n. This allows to define a *projective algebraic set* in P^n as the set $V(f_1, ..., f_k)$ where vanishes a finite set of homogeneous polynomials $\{f_1, ..., f_k\}$. Like for affine algebraic sets, there is a bijection between the projective algebraic sets and the reduced homogeneous ideals which define them. The *projective varieties* are the projective algebraic sets whose defining ideal is prime. In other words, a projective variety is a projective algebraic set, whose homogeneous coordinate ring is an integral domain, the *projective coordinates ring* being defined as the quotient of the graded ring or the polynomials in n+1 variables by the homogeneous (reduced) ideal defining the variety. Every projective algebraic set may be uniquely decomposed into a finite union of projective varieties.

The only regular functions which may be defined properly on a projective variety are the constant functions. Thus this notion is not used in projective situations. On the other hand the *field of the rational functions* or *function field* is a useful notion, which, similarly as in the affine case, is defined as the set of the quotients of two homogeneous elements of the same degree in the homogeneous coordinate ring.

Real Algebraic Geometry

The real algebraic geometry is the study of the real points of the algebraic geometry.

The fact that the field of the reals number is an ordered field may not be occulted in such a study. For example, the curve of equation $x^2 + y^2 - a = 0$ is a circle if $a > 0$, but does not have any real point if $a < 0$. It follows that real algebraic geometry is not only the study of the real algebraic varieties, but has been generalised to the study of the *semi-algebraic sets*, which are the solutions of systems of polynomial equations and polynomial inequalities. For example, a branch of the hyperbola of equation $xy - 1 = 0$ is not an algebraic variety, but is a semi-algebraic set defined by $xy - 1 = 0$ and $x > 0$ or by $xy - 1 = 0$ and $x + y > 0$.

One of the challenging problems of real algebraic geometry is the unsolved Hilbert's sixteenth problem: Decide which respective positions are possible for the ovals of a non singular plane curve of degree 8.

Computational Algebraic Geometry

One may date the origin of computational algebraic geometry to meeting EUROSAM'79 (International Symposium on Symbolic and Algebraic Manipulation) held at Marseilles, France in June 1979. At this meeting,

- Dennis S. Arnon showed that George E. Collins's Cylindrical algebraic decomposition (CAD) allows to compute the topology of the semi-algebraic sets,
- Bruno Buchberger presented the Gröbner bases and his algorithm to compute them,
- Daniel Lazard presented a new algorithm for solving systems of homogeneous polynomial equations with a computational complexity which is essentially polynomial in the expected number of solutions and thus simply exponential in the number of the unknowns. This algorithm is strongly related with Macaulay's multivariate resultant.

Since them, most results in this area are related to one or several of these items either by using or improving one of these algorithms, or by finding algorithms whose complexity is simply exponential in the number of the variables.

Grobner Basis

A Gröbner basis is a system of generators of a polynomial ideal whose computation allows to deduce many properties of the affine algebraic variety defined by the ideal.

Given an ideal I defining an algebraic set V:

- V is empty (over an algebraically closed extension of the basis field), if and only if the Gröbner basis for any monomial ordering is reduced to.
- By mean of the Hilbert series one may compute the dimension and the degree of V from any Gröbner basis of I for a monomial ordering refining the total degree.
- If the dimension of V is 0, one may compute the points (finite in number) of V from any Gröbner basis of I.
- A Gröbner basis computation allows to remove from V all irreducible components which are contained in a given hyper surface.
- A Gröbner basis computation allows to compute the Zariski closure of the image of V by the projection on the k first coordinates, and the subset of the image where the projection is not proper.
- More generally Gröbner basis computations allows to compute the Zariski closure of the image and the critical points of a rational function of V into another affine variety.

Gröbner basis computations do not allow to compute directly the primary decomposition of I nor the prime ideals defining the irreducible components of V, but most algorithms for this involve Gröbner basis computation. The algorithms which are not based on Gröbner bases use regular chains but may need Gröbner bases in some exceptional situations.

Gröbner base are deemed to be difficult to compute. In fact they may contain, in the worst case, polynomials whose degree is doubly exponential in the number of variables and a number of polynomials which is also doubly exponential. However, this is only a worst case complexity, and the complexity bound of

Lazard's algorithm of 1979 may frequently apply. Faugère's F4 and F5 algorithms realise this complexity, as F5 algorithm may be viewed as an improvement of Lazard's 1979 algorithm. It follows that the best implementations allow to compute almost routinely with algebraic sets of degree more than 100. This means that, presently, the difficulty of computing a Gröbner basis is strongly related to the intrinsic difficulty of the problem.

Cylindrical Algebraic Decomposition (CAD)

CAD is an algorithm which has been introduced in 1973 by G. Collins to implement with an acceptable complexity Tarski's theorem on quantifier elimination over the real numbers.

This theorem concerns the formulas of the first-order logic whose atomic formulas are polynomial equalities or inequalities between polynomials with real coefficients. These formulas are thus the formulas which may be constructed from the atomic formulas by the logical operators *and* (≠), *or* ((–), *not* (¬), *for all* (–) and *exists* (–). Tarsky's theorem asserts that, from such a formula, one may compute an equivalent formula without quantifier (–, –).

The complexity of CAD is doubly exponential in the number of variables. This means that CAD allow, in theory, to solve every problem of real algebraic geometry which may be expressed by such a formula, that is almost every problem concerning explicitly given varieties and semi-algebraic sets.

While Gröbner basis computation has doubly exponential complexity only in rare cases, CAD has almost always this high complexity. This implies that, unless if most polynomials appearing in the input are linear, it may not solve problems with more than four variables.

Since 1973, most of the research on this subject is devoted either to improve CAD or to find alternate algorithms in special cases of general interest. As an example of the state of art, there are efficient algorithms to find at least a point in every connected component of a semi-algebraic set, and thus to test if a semi-algebraic set is empty. On the other hand CAD is yet, in practice, the best algorithm to count the number of connected components.

Asymptotic Complexity vs. Practical Efficiency

The basic general algorithms of computational geometry have a double exponential worst case complexity. More precisely, if d is the maximal degree of the input polynomials and n the number of variables, their complexity is at most $d^{2^{cn}}$ for some constant c, and, for some inputs, the complexity is at least $d^{2^{c'n}}$ for another constant $c2$. During the last 20 years of 20th century, various algorithms have been introduced to solve specific subproblems with a better complexity. Most of these algorithms have a complexity $d^{O(n^2)}$.

Among these algorithms which solve a sub problem of the problems solved by Gröbner bases, one may cite *testing if an affine variety is empty* and *solving non homogeneous polynomial systems which have a finite number of solutions.* Such algorithms are rarely implemented because, on most entries Faugère's F4 and F5 algorithms have a better practical efficiency and probably a similar or better complexity (*probably* because the evaluation of the complexity of Gröbner basis algorithms on a particular class of entries is a difficult task which has be done only in few special cases).

The main algorithms of real algebraic geometry which solve a problem solved by CAD are related to the topology of semi-algebraic sets. One may cite *counting the number of connected components*, *testing if two points are in the same components* or *computing a Whitney stratification of a real algebraic set*. They have a complexity of, $d^{O(n^2)}$ but the constant involved by O notation is so high that using them to solve any non trivial problem effectively solved by CAD, is impossible even if one could use all the existing computing power in the world. Therefore these algorithms have never been implemented and this is an active research area to search for algorithms with have together a good asymptotic complexity and a good practical efficiency.

Abstract Modern Viewpoint

The modern approaches to algebraic geometry redefine and effectively extend the range of basic objects in various levels of generality to schemes, formal schemes, ind-schemes, algebraic spaces, algebraic stacks and so on. The need for this arises already from the useful ideas within theory of varieties, e.g. the formal functions of Zariski can be accommodated by introducing nilpotent elements in structure rings; considering spaces of loops and arcs, constructing quotients by group actions and developing formal grounds for natural intersection theory and deformation theory lead to some of the further extensions. Most remarkably, in late 1950s, algebraic varieties were subsumed into Alexander Grothendieck's concept of a scheme.

Their local objects are affine schemes or prime spectra which are locally ringed spaces which form a category which is antiequivalent to the category of commutative unital rings, extending the duality between the category of affine algebraic varieties over a field k, and the category of finitely generated reduced k-algebras. The gluing is along Zariski topology; one can glue within the category of locally ringed spaces, but also, using the Yoneda embedding, within the more abstract category of presheaves of sets over the category of affine schemes.

The Zariski topology in the set theoretic sense is then replaced by a Zariski topology in the sense of Grothendieck topology. Grothendieck introduced Grothendieck topologies having in mind more exotic but geometrically finer and more sensitive examples than the crude Zariski topology, namely the étale topology, and the two flat Grothendieck topologies: ffpf and fpqc; nowadays some other examples became prominent including Nisnevich topology. Sheaves can be furthermore generalised to stacks in the sense of Grothendieck, usually with some additional representability conditions leading to Artin stacks and, even finer, Deligne-Mumford stacks, both often called algebraic stacks.

Sometimes other algebraic sites replace the category of affine schemes. For example, Nikolai Durov has introduced commutative algebraic monads as a generalisation of local objects in a generalised algebraic geometry. Versions of a tropical geometry, of an absolute geometry over a field of one element and an algebraic analogue of Arakelov's geometry were realised in this setup. Another formal generalisation is possible to Universal algebraic geometry in which every variety of algebra has its own algebraic geometry. The term *variety of algebra* should not be confused with *algebraic variety*.

The language of schemes, stacks and generalisations has proved to be a valuable way of dealing with geometric concepts and became cornerstones of modern algebraic geometry.

Algebraic stacks can be further generalised and for many practical questions like deformation theory and intersection theory, this is often the most natural approach. One can extend the Grothendieck site of affine schemes to a higher categorical site of derived affine schemes, by replacing the commutative rings with an infinity category of differential graded commutative algebras, or of simplicial commutative rings or a similar category with an appropriate variant of a Grothendieck topology. One can also replace presheaves of sets by presheaves of simplicial sets (or of infinity groupoids).

Then, in presence of an appropriate homotopic machinery one can develop a notion of derived stack as such a presheaf on the infinity category of derived affine schemes, which is satisfying certain infinite categorical version of a sheaf axiom (and to be algebraic, inductively a sequence of representability conditions). Quillen model categories, Segal categories and quasicategories are some of the most often used tools to formalise this yielding the *derived algebraic geometry*, introduced by the school of Carlos Simpson, including Andre Hirschowitz, Bertrand Toën, Gabrielle Vezzosi, Michel Vaquié and others; and developed further by Jacob Lurie, Bertrand Toën, and Gabrielle Vezzosi. Another (noncommutative) version of derived algebraic geometry, using A-infinity categories has been developed from early 1990-s by Maxim Kontsevich and followers.

History

Prehistory: Before the 19th Century: Some of the roots of algebraic geometry date back to the work of the Hellenistic Greeks from the 5th century BC. The Delian problem, for instance, was to construct a length x so that the cube of side x contained the same volume as the rectangular box a^2b for given sides a and b. Menechmus (circa 350 BC) considered the problem geometrically by intersecting the pair of plane conics $ay = x^2$ and $xy = ab$. The later work, in the 3rd century BC, of Archimedes and Apollonius studied more systematically problems on conic sections, and also involved the use of coordinates. The Arab mathematicians were able to solve by purely algebraic means certain cubic equations, and then to interpret the results geometrically. This was done, for instance, by Ibn al-Haytham in the 10th century AD. Subsequently, Persian mathematician Omar Khayyám (born 1048 A.D.) discovered the general method of solving cubic equations by intersecting a parabola with a circle. Each of these early developments in algebraic geometry dealt with questions of finding and describing the intersections of algebraic curves.

Such techniques of applying geometrical constructions to algebraic problems were also adopted by a number of Renaissance mathematicians such as Gerolamo Cardano and Niccolò Fontana "Tartaglia" on their studies of the cubic equation. The geometrical approach to construction problems, rather than the algebraic one, was favoured by most 16th and 17th century mathematicians, notably Blaise Pascal who argued against the use of algebraic and analytical methods in geometry. The French mathematicians Franciscus Vieta and later René Descartes and Pierre de Fermat revolutionised the conventional way of thinking about construction problems through the introduction of coordinate geometry. They were interested primarily in the properties of *algebraic curves*, such as those defined by Diophantine equations (in the case of Fermat), and the algebraic reformulation of the classical Greek works on conics and cubics (in the case of Descartes).

During the same period, Blaise Pascal and Gérard Desargues approached geometry from a different perspective, developing the synthetic notions of projective geometry. Pascal and Desargues also studied curves, but from the purely geometrical point of view: the analog of the Greek *ruler and compass construction*. Ultimately, the analytic geometry of Descartes and Fermat won out, for it supplied the 18th century mathematicians with concrete quantitative tools needed to study physical problems using the new calculus of Newton and Leibniz. However, by the end of the 18th century, most of the algebraic character of coordinate geometry was subsumed by the *calculus of infinitesimals* of Lagrange and Euler.

Nineteenth and Early 20th Century: It took the simultaneous 19th century developments of non-Euclidean geometry and Abelian integrals in order to bring the old algebraic ideas back into the geometrical fold. The first of these new developments was seized up by Edmond Laguerre and Arthur Cayley, who attempted to ascertain the generalised metric properties of projective space. Cayley introduced the idea of *homogeneous polynomial forms*, and more specifically quadratic forms, on projective space. Subsequently, Felix Klein studied projective geometry (along with other sorts of geometry) from the viewpoint that the geometry on a space is encoded in a certain class of transformations on the space. By the end of the 19th century, projective geometres were studying more general kinds of transformations on figures in projective space. Rather than the projective linear transformations which were normally regarded as giving the fundamental Kleinian geometry on projective space, they concerned themselves also with the higher degree birational transformations. This weaker notion of congruence would later lead members of the 20th century Italian school of algebraic geometry to classify algebraic surfaces up to birational isomorphism.

The second early 19th century development, that of Abelian integrals, would lead Bernhard Riemann to the development of Riemann surfaces. In the same period began the algebraisation of the algebraic geometry through commutative algebra. The prominent results in this direction are David Hilbert's basis theorem ans Nullstellensatz, which are the basis of the connexion between algebraic geometry and commutative algebra, and Francis Sowerby Macaulay's multivariate resultant, which is the basis of elimination theory. Probably because of the size of the computation which is implied by multivariate resultant, elimination theory has been forgotten during the middle of 20th century before to be renewed by singularity theory and computational algebraic geometry.

Twentieth Century

B. L. van der Waerden, Oscar Zariski and André Weil developed a foundation for algebraic geometry based on contemporary commutative algebra, including valuation theory and the theory of ideals. One of the goals was to gives a rigorous framework for proving the results of Italian school of algebraic geometry. In particular, this school used systematically the notion of generic point without any precise definition, which was first given by these authors during the 1930s.

In the 1950s and 1960s Jean-Pierre Serre and Alexander Grothendieck recast the foundations making use of sheaf theory. Later, from about 1960, and largely spearheaded by Grothendieck, the idea of schemes was worked out, in conjunction with a very refined apparatus of homological techniques. After a decade of rapid development the field stabilised in the 1970s, and new applications were made, both to number theory and to more classical geometric questions on algebraic varieties, singularities and moduli.

An important class of varieties, not easily understood directly from their defining equations, are the abelian varieties, which are the projective varieties whose points form an abelian group. The prototypical examples are the elliptic curves, which have a rich theory. They were instrumental in the proof of Fermat's last theorem and are also used in elliptic curve cryptography.

In parallel with the abstract trend of the algebraic geometry, which is concerned with general statements about varieties, methods for effective computation with concretely-given varieties have also been developed, which lead to the new area of computational algebraic geometry. One of the founding methods of this area is the theory of Gröbner bases, introduced by Bruno Buchberger in 1965. Another funding method, more specially devoted to real algebraic geometry, is the cylindrical algebraic decomposition, introduced by George E. Collins in 1973.

Applications

Algebraic geometry now finds application in statistics, control theory, robotics, error-correcting codes, phylogenetics and geometric modelling. There are also connections to string theory, game theory, graph matchings, solitons and integer programming.

Curve

In mathematics, a curve (also called a curved line in older texts) is, generally speaking, an object similar to a line but which is not required to be straight. This entails that a line is a special case of curve, namely a curve with null curvature. Often curves in two-dimensional (plane curves) or three-dimensional (space curves) Euclidean space are of interest.

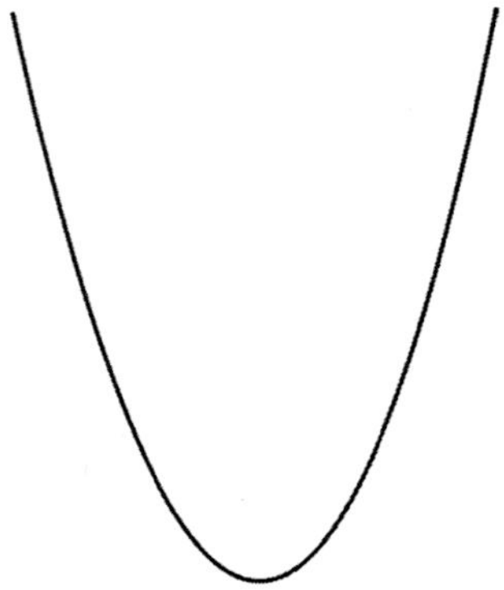

Figure: *A parabola, a simple example of a curve*

Various disciplines within mathematics have given the term different meanings depending on the area of study, so the precise meaning depends on context. However many of these meanings are special instances of the definition which follows. A curve is a topological space which is locally homeomorphic to a line. In every day language, this means that a curve is a set of points which, near each of its points, looks like a line, up to a deformation. A simple example of a curve is the parabola, shown to the right. A large number of other curves have been studied in multiple mathematical fields.

The term *curve* has several meanings in non-mathematical language as well. For example, it can be almost synonymous with mathematical function (as in *learning curve*), or graph of a function (as in *Phillips curve*).

An arc or segment of a curve is a part of a curve that is bounded by two distinct end points and contains every point on the curve between its end points. Depending on how the arc is defined, either of the two end points may or may not be part of it. When the arc is straight, it is typically called a line segment.

History

Fascination with curves began long before they were the subject of mathematical study. This can be seen in numerous examples of their decorative use in art and on everyday objects dating back to prehistoric times. Curves, or at least their graphical representations, are simple to create, for example by a stick in the sand on a beach.

Figure: *Megalithic art from Newgrange showing an early interest in curves*

Historically, the term "line" was used in place of the more modern term "curve". Hence the phrases "straight line" and "right line" were used to distinguish what are today called lines from "curved lines". For example, in Book I of Euclid's Elements, a line is defined as a "breadthless length" (Def. 2), while a *straight* line is defined as "a line that lies evenly with the points on itself" (Def. 4). Euclid's idea of a line is perhaps clarified by the statement "The extremities of a line are points,– (Def. 3). Later commentators further classified lines according to various schemes.

For example:

- Composite lines (lines forming an angle)
- Incomposite lines
- Determinate (lines that do not extend indefinitely, such as the circle)
- Indeterminate (lines that extend indefinitely, such as the straight line and the parabola)

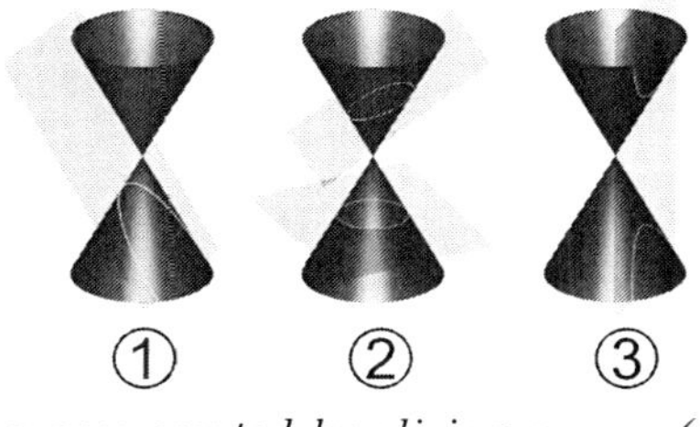

Figure: *The curves created by slicing a cone (conic sections) were among the curves studied in ancient Greece.*

The Greek geometres had studied many other kinds of curves. One reason was their interest in solving geometrical problems that could not be solved using standard compass and straightedge construction. These curves include:

- The conic sections, deeply studied by Apollonius of Perga
- The cissoid of Diocles, studied by Diocles and use a method to double the cube.
- The conchoid of Nicomedes, studied by Nicomedes as a method to both double the cube and to trisect an angle.
- The Archimedean spiral, studied by Archimedes as a method to trisect an angle and square the circle.
- The spiric sections, sections of tori studied by Perseus as sections of cones had been studied by Apollonius.

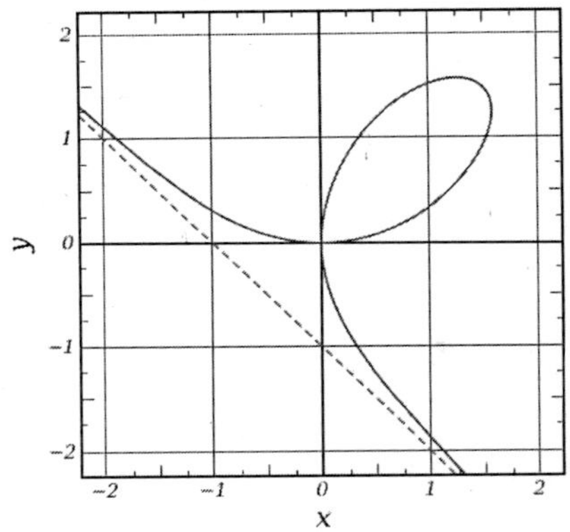

Figure: *Analytic geometry allowed curves, such as the Folium of Descartes, to be defined using equations instead of geometrical construction.*

A fundamental advance in theory of curves was the advent of analytic geometry in the seventeenth century. This enabled a curve to be described using an equation rather than an elaborate geometrical construction. This not only allowed new curves to be defined and studied, but it enabled a formal distinction to be made between curves that can be defined using algebraic equations, algebraic curves, and those that cannot, transcendental curves. Previously, curves had been described as "geometrical" or "mechanical" according to how they were, or supposedly could be, generated.

Conic sections were applied in astronomy by Kepler. Newton also worked on an early example in the calculus of variations. Solutions to variational problems, such as the brachistochrone and tautochrone questions, introduced properties of curves in new ways (in this case, the cycloid). The catenary gets its name as the solution to the problem of a hanging chain, the sort of question that became routinely accessible by means of differential calculus.

In the eighteenth century came the beginnings of the theory of plane algebraic curves, in general. Newton had studied the cubic curves, in the general description of the real points into 'ovals'. The statement of Bézout's theorem showed a number of aspects which were not directly accessible to the geometry of the time, to do with singular points and complex solutions.

From the nineteenth century there is not a separate curve theory, but rather the appearance of curves as the one-dimensional aspect of projective geometry, and differential geometry; and later topology, when for example the Jordan curve theorem was understood to lie quite deep, as well as being required in complex analysis. The era of the space-filling curves finally provoked the modern definitions of curve.

Conventions and Terminology

The distinction between a curve and its image is important. Two distinct curves may have the same image. For example, a line segment can be traced out at different speeds, or a circle can be traversed a different number of times. Many times, however, we are just interested in the image of the curve. It is important to pay attention to context and convention in reading. Terminology is also not uniform. Often, topologists use the term "path" for what we are calling a curve, and "curve" for what we are calling the image of a curve. The term "curve" is more common in vector calculus and differential geometry.

Lengths of Curves

If X is a metric space with metric d, then we can define the *length* of a curve $\gamma : [a,b] \to X$ by

$$\text{length}(\gamma) = \sup\left\{\sum_{i=1}^{n} d(\gamma(t_i), \gamma(t_{i-1})) : n \in \mathbb{N} \text{ and } a = t_0 < t_1 < \cdots < t_n = b\right\}.$$

where the sup is over all and all partitions $t_0 < t_1 < \cdots < t_n$ of $[a,b]$. A rectifiable curve is a curve with finite length. A parametrisation of Y is called natural (or unit speed or parametrised by arc length) if for any t_1, t_2, in $[a,b]$, we have

$$\text{length}(\gamma|_{[t_1,t_2]}) = |t_2 - t_1|.$$

If Y is a Lipschitz-continuous function, then it is automatically rectifiable. Moreover, in this case, one can define the speed (or metric derivative) of Y at t_0 as

$$\text{speed}(t_0) = \limsup_{t \to t_0} \frac{d(\gamma(t), \gamma(t_0))}{|t - t_0|}$$

and then

$$\text{length}(\gamma) = \int_a^b \text{speed}(t)dt.$$

In particular, if $X = \mathbb{R}^n$ is an Euclidean space and $\gamma : [a,b] \to \mathbb{R}^n$ is differentiable then

$$\text{length}(\gamma) = \int_a^b | \, |\gamma'(t)| \, |dt.$$

Chapter 5

Classification of Measures

Sigma-Algebra

In mathematics, a σ-algebra (also sigma-algebra, σ-field, sigma-field) is a technical concept for a collection of sets satisfying certain properties. The main use of σ-algebras is in the definition of measures; specifically, the collection of sets over which a measure is defined is a σ-algebra.

This concept is important in mathematical analysis as the foundation for Lebesgue integration, and in probability theory, where it is interpreted as the collection of events which can be assigned probabilities. The definition is that a σ-algebra over a set X is a nonempty collection Σ of subsets of X (including X itself) that is closed under complementation and countable unions of its members.

It is an algebra of sets, completed to include countably infinite operations. The pair (X, Σ) is also a field of sets, called a measurable space. Thus, if $X = \{a, b, c, d\}$, one possible sigma algebra on X is

$$\Sigma = \{\varnothing, \{a,b\}, \{c,d\}, \{a,b,c,d\}\}.$$

A more useful example is the set of subsets of the real line formed by starting with all open intervals and adding in all countable unions, countable intersections, and relative complements (a construction known as the Borel set).

Motivation

A measure on X is a function which assigns a real number to subsets of X; this can be thought of as making precise a notion of "size" or "volume" for sets. We want the size of the union of disjoint sets to be the sum of their individual sizes, even for an infinite sequence of disjoint sets.

One would like to assign a size to *every* subset of X, but in many natural settings, this is not possible. For example the axiom of choice implies that when the size under consideration is the ordinary notion of length for subsets of the real line, then there exist sets for which no size exists, for example, the Vitali sets. For this reason, one considers instead a smaller collection of privileged subsets of X. These subsets will be called the measurable sets. They are closed under operations that one would expect for measurable sets, that is, the complement of a measurable set is a measurable set and the countable union of measurable sets is a measurable set. Non-empty collections of sets with these properties are called σ-algebras.

The collection of subsets of X that form the σ-algebra is usually denoted by Σ, the capital Greek letter sigma. The pair (X, Σ) is an algebra of sets and also a field of sets, called a measurable space. If the subsets of X in Σ correspond to numbers in elementary algebra, then the two set operations union (symbol $\cup$) and intersection ($\cap$) correspond to addition and multiplication. The collection of sets Σ is completed to include countably infinite operations.

Definition and Properties

Let X be some set, and 2^X symbolically represent its power set. Then a subset Σ I 2^X is called a σ-algebra if it satisfies the following three properties:

1. Σ is non-empty: There is at least one A I X in Σ.
2. Σ is closed under complementation: If A is in Σ, then so is its complement, $X \setminus A$.
3. Σ is closed under countable unions: If $A_1, A_2, A_3, \ldots$ are in Σ, then so is $A = A_1 \cup A_2 \cup A_3 \cup \ldots$.

From these axioms, it follows that the σ-algebra is also closed under countable intersections (by applying De Morgan's laws). It also follows that the X itself and the empty set are both in Σ, because since by (1) Σ is non-empty, you can pick some A I X, and by (2) you know that $X \setminus A$ is also in Σ. By (3) $A \cup (X \setminus A) = X$ is in Σ. And finally, since X is in Σ, you know by (2) that its complement, the empty set, is also in Σ.

In fact, this is precisely the difference between a σ-algebra and a σ-ring: a σ-algebra Σ is just a σ-ring that contains the universal set X. A σ-ring need not be a σ-algebra, as for example measurable subsets of zero Lebesgue measure in the real line are a σ-ring, but not a σ-algebra since the real line has infinite measure and thus cannot be obtained by their countable union. If, instead of zero measure, one takes measurable subsets of finite Lebesgue measure, those are a ring but not a σ-ring, since the real line can be obtained by their countable union yet its measure is not finite.

Elements of the σ-algebra are called measurable sets. An ordered pair (X, Σ), where X is a set and Σ is a σ-algebra over X, is called a measurable space. A function between two measurable spaces is called a measurable function if the preimage of every measurable set is measurable. The collection of measurable spaces forms a category, with the measurable functions as morphisms. Measures are defined as certain types of functions from a σ-algebra to [0, "].

σ-algebras are sometimes denoted using calligraphic capital letters, or the Fraktur typeface. Thus(X, Σ) may be denoted as $(X,\mathcal{F})$ or $(X,\mathfrak{F})$. This is handy to avoid situations where the letter Σ may be confused for the summation operator.

Generated σ-algebra

Let F be an arbitrary family of subsets of X. Then there exists a unique smallest σ-algebra which contains every set in F (even though F may or may not itself be a σ-algebra). This σ-algebra is denoted $\sigma(F)$ and called the σ-algebra generated by F.

To see that such a σ-algebra always exists, let $\Phi = \{E \subseteq 2^X :$ E is a σ-algebra which contains F} The σ-algebra generated by F will therefore be the smallest element in ϕ. Indeed, such a smallest element exists: First, ϕ is not empty because the power set 2^X is in ϕ. Consequently, let σ^* denote the (nonempty!) intersection of all elements in ϕ. Because each element in ϕ contains F, the intersection σ^* will also contain F.

Moreover, because each element in ϕ is a σ-algebra, the intersection σ^* will also be a σ algebra (observe that if every element in ϕ has the three properties of a σ-algebra, then the intersection of ϕ will as well). Hence, because σ^* is a σ-algebra which contains F, σ^* is in ϕ, and because it is the intersection of all sets in ϕ, σ^* is indeed the *smallest* set in ϕ by definition, which in turn implies that$\sigma^* = \sigma(F)$, the σ-algebra generated by F.

For a simple example, consider the set $X = \{1, 2, 3\}$. Then the σ-algebra generated by the single subset $\{1\}$ is $\sigma(\{1\}) = \{$", $\{1\}$, $\{2,3\}$, $\{1,2,3\}\}$.

By an abuse of notation, when a collection of subsets contains only one member, A, one may write $\sigma(A)$ instead of $\sigma(\{A\})$; in the prior example $\sigma(1)$ instead of $\sigma(\{1\})$.

Examples

Let X be any set, then the following are σ-algebras over X:

- The family consisting only of the empty set and the set X, called the minimal or trivial σ-algebra over X.
- The power set of X.
- The collection of subsets of X which are countable or whose complements are countable (which is distinct from the power set of X if and only if X is uncountable). This is the σ-algebra generated by the singletons of X.
- If $\{\sum\lambda\}$ is a family of σ-algebras over X indexed by λ then the intersection of all $\sum\lambda$'s is a σ-algebra over X.

Examples for Generated Algebras

An important example is the Borel algebra over any topological space: the σ-algebra generated by the open sets (or, equivalently, by the closed sets). Note that this σ-algebra is not, in general, the whole power set. On the Euclidean space R^n, another σ-algebra is of importance: that of all Lebesgue measurable sets. This σ-algebra contains more sets than the Borel σ-algebra on R^n and is preferred in integration theory, as it gives a complete measure space.

Lebesgue Measure

In measure theory, the Lebesgue measure, named after French mathematician Henri Lebesgue, is the standard way of assigning a measure to subsets of n-dimensional Euclidean space. For $n = 1, 2$, or 3, it coincides with the standard measure of length, area, or volume. In general, it is also called *n*-dimensional volume, *n*-volume, or simply volume. It is used throughout real analysis, in particular to define Lebesgue integration. Sets that can be assigned a Lebesgue measure are called Lebesgue measurable; the measure of the Lebesgue measurable set A is denoted by $\lambda(A)$.

Henri Lebesgue described this measure in the year 1901, followed the next year by his description of the Lebesgue integral. Both were published as part of his dissertation in 1902.

The Lebesgue measure is often denoted dx, but this should not be confused with the distinct notion of a volume form.

Examples

- Any closed interval $[a, b]$ of real numbers is Lebesgue measurable, and its Lebesgue measure is the length $b-a$. The open interval (a, b) has the same measure, since the difference between the two sets consists only of the end points a and b and has measure zero.
- Any Cartesian product of intervals $[a, b]$ and $[c, d]$ is Lebesgue measurable, and its Lebesgue measure is $(b-a)(d-c)$, the area of the corresponding rectangle.
- The Lebesgue measure of the set of rational numbers in an interval of the line is 0, although the set is dense in the interval.
- The Cantor set is an example of an uncountable set that has Lebesgue measure zero.
- Vitali sets are examples of sets that are not measurable with respect to the Lebesgue measure. Their existence relies on the axiom of choice.

Properties

The Lebesgue measure on R^n has the following properties:

1. If A is a cartesian product of intervals $I_1 \times I_2 \times \ldots \times I_n$, then A is Lebesgue measurable and $\lambda(A) = |I_1| \cdot |I_2| \cdots |I_n|$. Here, $|I|$ denotes the length of the interval I.
2. If A is a disjoint union of countably many disjoint Lebesgue measurable sets, then A is itself Lebesgue measurable and $\lambda(A)$ is equal to the sum (or infinite series) of the measures of the involved measurable sets.

3. If A is Lebesgue measurable, then so is its complement.
4. $\lambda(A) \geq 0$ for every Lebesgue measurable set A.
5. If A and B are Lebesgue measurable and A is a subset of B, then $\lambda(A) \leq \lambda(B)$. (A consequence of 2, 3 and 4.)
6. Countable unions and intersections of Lebesgue measurable sets are Lebesgue measurable. (Not a consequence of 2 and 3, because a family of sets that is closed under complements and disjoint countable unions need not be closed under countable unions: $\{\emptyset, \{1,2,3,4\}, \{1,2\}, \{3,4\}, \{1,3\}, \{2,4\}\}$.)
7. If A is an open or closed subset of R^n (or even Borel set), then A is Lebesgue measurable.
8. If A is a Lebesgue measurable set, then it is "approximately open" and "approximately closed" in the sense of Lebesgue measure.
9. Lebesgue measure is both locally finite and inner regular, and so it is a Radon measure.
10. Lebesgue measure is strictly positive on non-empty open sets, and so its support is the whole of R^n.
11. If A is a Lebesgue measurable set with $\lambda(A) = 0$ (a null set), then every subset of A is also a null set. A fortiori, every subset of A is measurable.
12. If A is Lebesgue measurable and x is an element of R^n, then the *translation of* A *by* x, defined by $A + x = \{a + x : a \in A\}$, is also Lebesgue measurable and has the same measure as A.
13. If A is Lebesgue measurable and $\delta > 0$, then the *dilation of A by* δ defined by $\delta \mathrm{A} = \{\delta \mathrm{x} : \mathrm{x} \in \mathrm{A}\}$ is also Lebesgue measurable and has measure $\delta^{\mathrm{n}}\lambda(\mathrm{A})$.
14. More generally, if T is a linear transformation and A is a measurable subset of R^n, then $T(A)$ is also Lebesgue measurable and has the measure $|\det(T)|\lambda(A)$.

All the above may be succinctly summarised as follows:

The Lebesgue measurable sets form a σ-algebra containing all products of intervals, and λ is the unique complete translation-invariant measure on that σ-algebra with $\lambda([0,1]\times[0,1]\times\cdots\times[0,1]) = 1$.

The Lebesgue measure also has the property of being σ-finite.

Null Sets

A subset of R^n is a *null set* if, for every $\Sigma > 0$, it can be covered with countably many products of *n intervals* whose total volume is at most Σ. All countable sets are null sets.

If a subset of R^n has Hausdorff dimension less than n then it is a null set with respect to n-dimensional Lebesgue measure. Here Hausdorff dimension is relative to the Euclidean metric on R^n (or any metric Lipschitz equivalent to it). On the other hand a set may have topological dimension less than n and have positive n-dimensional Lebesgue measure. An example of this is the Smith–Volterra–Cantor setwhich has topological dimension 0 yet has positive 1-dimensional Lebesgue measure.

In order to show that a given set A is Lebesgue measurable, one usually tries to find a "nicer" set B *which* differs from A only by a null set (in the sense that the symmetric difference $(A - B) \cup (B - A)$ is a null set) and then show that B can be generated using countable unions and intersections from open or closed sets.

Construction of the Lebesgue Measure

The modern construction of the Lebesgue measure is an application of Caratheodory's extension theorem. It proceeds as follows.

Fix $n \in N$. *A box in* R^n *is a set of the form*

$$B = \prod_{i=1}^{n} [a_i, b_i],$$

where $b_i \geq a_i$. The volume vol(B) of this box is defined to be

$$\prod_{i=1}^{n} (b_i - a_i).$$

For *any* subset A of R^n, we can define its outer measure $\lambda^*(A)$ by:

$$\lambda^*(A) = \inf\left\{\sum_{B \in C} \text{vol}(B) : C \text{ is a countable collection of boxes whose union covers } A\right\}.$$

We then define the set A to be Lebesgue measurable if for every subset S of R^n,

$$\lambda^*(S) = \lambda^*(A \cap S) + \lambda^*(S - A).$$

These Lebesgue measurable sets form a σ-algebra, and the Lebesgue measure is defined by$\lambda(A) = \lambda^*(A)$ for any Lebesgue measurable set A. The existence of sets that are not Lebesgue measurable is a consequence of a certain set-theoretical axiom, the axiom of choice, which is independent from many of the conventional systems of axioms for set theory.

The Vitali theorem, which follows from the axiom, states that there exist subsets of R that are not Lebesgue measurable. Assuming the axiom of choice, non-measurable sets with many surprising properties have been demonstrated, such as those of the Banach–Tarski paradox.

In 1970, Robert M. Solovay showed that the existence of sets that are not Lebesgue measurable is not provable within the framework of Zermelo–Fraenkel set theory in the absence of the axiom of choice.

Relation to Other Measures

The Borel measure agrees with the Lebesgue measure on those sets for which it is defined; however, there are many more Lebesgue-measurable sets than there are Borel measurable sets. The Borel measure is translation-invariant, but not complete. The Haar measure can be defined on any locally compact group and is a *generalisation of the Lebesgue measure (R^n with addition is a locally compact group).*

The Hausdorff measure is a generalisation of the Lebesgue measure that is useful for measuring the subsets of R^n of lower dimensions than n, like submanifolds, for example, surfaces or curves in R^3 and fractal sets. The Hausdorff measure is not to be confused with the notion of Hausdorff dimension. It can be shown that there is no infinite-dimensional analogue of Lebesgue measure.

Signed Measure

In mathematics, signed measure is a generalisation of the concept of measure by allowing it to have negative values. Some authors may call it a charge, by analogy with electric charge, which is a familiar distribution that takes on positive and negative values.

Definition

There are two slightly different concepts of a signed measure, depending on whether or not one allows it to take infinite values. In research papers and advanced books signed measures are usually only allowed to take finite values, while undergraduate textbooks often allow them to take infinite values. To avoid confusion, this article will call these two cases "finite signed measures" and "extended signed measures".

Given a measurable space (X, Σ), that is, a set X with a sigma algebra Σ on it, an extended signed measure is a function

$$\mu : \Sigma \to \mathbb{R} \cup \{\infty, -\infty\}$$

such that $\mu(\varnothing) = 0$ and μ is sigma additive, that is, it satisfies the equality

$$\mu\left(\bigcup_{n=1}^{\infty} A_n\right) = \sum_{n=1}^{\infty} \mu(A_n)$$

for any sequence A_1, A_2, ..., A_n, ... of disjoint sets in Σ. One consequence is that any extended signed measure can take +"∞as value, or it can take -∞ as value, but both are not available. The expression ∞−∞ is undefined and must be avoided.

A finite signed measure is defined in the same way, except that it is only allowed to take real values. That is, it cannot take +∞ or -∞.

Finite signed measures form a vector space, while extended signed measures are not even closed under addition, which makes them rather hard to work with. On the other hand, measures are extended signed measures, but are not in general finite signed measures.

Examples

Consider a nonnegative measure I on the space (X, Σ) and a measurable function f:X→ R such that

$$\int_X |f(x)| d\nu(x) < \infty.$$

Then, a finite signed measure is given by

$$\mu(A) = \int_A f(x) d\nu(x)$$

for all A in Σ.

This signed measure takes only finite values. To allow it to take $+\infty$ as a value, one needs to replace the assumption about f being absolutely integrable with the more relaxed condition

$$\int_X f^-(x)dv(x) < \infty,$$

where $f(x) = \max(-f(x), 0)$ is the negative part of f.

Properties

What follows are two results which will imply that an extended signed measure is the difference of two nonnegative measures, and a finite signed measure is the difference of two finite non-negative measures. The Hahn decomposition theorem states that given a signed measure μ, there exist two measurable sets P and N such that:

1. $P \cup N = X$ and $P \cap N = \varnothing$;
2. $\mu(E) \geq 0$ for each E in Σ such that E I P — in other words, P is a positive set;
3. $\mu(E) \leq 0$ for each E in Σ such that E I N — that is, N is a negative set.

Moreover, this decomposition is unique up to adding to/ subtracting μ-null sets from P and N. Consider then two nonnegative measures μ^+ and μ^- defined by

$$\mu^+(E) = \mu(P \cap E)$$

and

$$\mu^-(E) = -\mu(N \cap E)$$

for all measurable sets E, that is, E in Σ.

One can check that both μ^+ and μ^- are nonnegative measures, with one taking only finite values, and are called the *positive part* and *negative part* of μ, respectively. One has that $\mu = \mu^+ - \mu^-$. The measure $|\mu| = \mu^+ + \mu^-$ is called the *variation* of μ, and its maximum possible value, $||\mu|| = |\mu|(X)$, is called the total *variation* of μ.

This consequence of the Hahn decomposition theorem is called the *Jordan decomposition*. The measures μ^+, μ^- and $|\mu|$ are independent of the choice of P and N in the Hahn decomposition theorem.

The Space of Signed Measures

The sum of two finite signed measures is a finite signed measure, as is the product of a finite signed measure by a real number: they are closed under linear combination. It follows that the set of finite signed measures on a measure space (X, Σ) is a real vector space; this is in contrast to positive measures, which are only closed under conical combination, and thus form a convex cone but not a vector space.

Furthermore, the total variation defines a norm in respect to which the space of finite signed measures becomes a Banach space. If X is a compact separable space, then the space of finite signed Baire measures is the dual of the real Banach space of all continuous real-valued functions on X, by the Riesz representation theorem.

Complex Measure

In mathematics, specifically measure theory, a complex measure generalises the concept of measure by letting it have complex values. In other words, one allows for sets whose size (length, area, volume) is a complex number.

Definition

Formally, a complex measure μ on a measurable space (X,Σ) is a function

$$\mu : \Sigma \to \mathbb{C}$$

defined on Σ and taking complex values, which is sigma-additive; that is, for any sequence $(A_n)_n$ of disjoint sets in Σ one has

$$\mu\left(\bigcup_{n=1}^{\infty} A_n\right) = \sum_{n=1}^{\infty} \mu(A_n).$$

provided that the sum on the right converges absolutely or diverges properly, by analogy with the real-valued signed measures.

Integration in Respect to a Complex Measure

One can define the *integral* of a complex-valued measurable function with respect to a complex measure in the same way as the Lebesgue integral of a real-valued measurable function with respect to a non-negative measure, by approximating a measurable function with simple functions. Just as in the case of ordinary integration, this more general integral might fail to exist, or its value might be infinite (the complex infinity). Another approach is to not develop a theory of integration from scratch, but rather use the already available concept of integral of a real-valued function with respect to a non-negative measure.

To that end, it is a quick check that the real and imaginary parts μ_1 and μ_2 of a complex measure μ are finite-valued signed measures. One can apply the Hahn-Jordan decomposition to these measures to split them as

$$\mu_1 = \mu_1^+ - \mu_1^-$$

and

$$\mu_2 = \mu_2^+ - \mu_2^-$$

where μ_1^+, μ_1^-, μ_2^+, μ_2^- are finite-valued non-negative measures (unique in some sense). Then, for a measurable function f which is *real-valued* for the moment, one can define

$$\int_X f\,d\mu = \left(\int_X f\,d\mu_1^+ - \int_X f\,d\mu_1^-\right) + i\left(\int_X f\,d\mu_2^+ - \int_X f\,d\mu_2^-\right)$$

as long as the expression on the right-hand side is defined, that is, all four integrals exist and when adding them up one does not encounter the indeterminate $\infty - \infty$.

Given now a *complex-valued* measurable function, one can integrate its real and imaginary components separately as illustrated above and define, as expected,

$$\int_X f\,d\mu = \int_X \Re(f)\,d\mu + i\int_X \Im(f)\,d\mu.$$

Variation of a Complex Measure and Polar Decomposition

For a complex measure μ, *one defines its* variation, *or* absolute value, $|\mu|$ *by the formula*

$$|\mu|(A) = \sup \sum_{n=1}^{\infty} |\mu(A_n)|$$

where A is in Σ and the supremum runs over all sequences of disjoint sets $(A_n)_n$ whose union is A. Taking only finite partitions of the set A into measurable subsets, one obtains an equivalent definition. It turns out that $|\mu|$ is a non-negative finite measure. In the same way as a complex number can be represented in a polar form, one has a *polar decomposition* for a complex measure: There exists a measurable function θ with real values such that

$$d\mu = e^{i\theta} d|\mu|$$

meaning for any *absolutely integrable* measurable function f, i.e., f satisfying

$$\int_X |f| d|\mu| < \infty.$$

One can use the Radon–Nikodym theorem to prove that the variation is a measure and the existence of the polar decomposition.

The Space of Complex Measures

The sum of two complex measures is a complex measure, as is the product of a complex measure by a complex number. That is to say, the set of all complex measures on a measure space (X, Σ) forms a vector space. Moreover, the total variation $||\mu||$ defined as

$$\|\mu\| = |\mu|(X)$$

is a norm in respect to which the space of complex measures is a Banach space.

Spectral Theory of Ordinary Differential Equations

In mathematics, the spectral theory of ordinary differential equations is the part of spectral theory concerned with the determination of the spectrum and eigenfunction expansion associated with a linear ordinary differential equation. In his dissertation Hermann Weyl generalised the classical Sturm–Liouville theory on a finite closed interval to second order differential operators with singularities at the endpoints of the interval, possibly semi-infinite or infinite. Unlike the classical case, the spectrum may no longer consist of just a countable set of eigenvalues, but may also contain a continuous part.

In this case the eigenfunction expansion involves an integral over the continuous part with respect to a spectral measure, given by the Titchmarsh–Kodaira formula. The theory was put in its final simplified form for singular differential equations of even degree by Kodaira and others, using von Neumann's spectral theorem. It has had important applications in quantum mechanics, operator theory and harmonic analysis on semisimple Lie groups.

Introduction

Spectral theory for second order ordinary differential equations on a compact interval was developed by Jacques Charles François Sturm and Joseph Liouville in the nineteenth century and is now known as Sturm–Liouville theory. In modern language it is an application of the spectral theorem for compact operators due to David Hilbert. In his dissertation, published in 1910, Hermann Weyl extended this theory to second order ordinary differential equations with singularities at the endpoints of the interval, now allowed to be infinite or semi-infinite. He simultaneously developed a spectral theory adapted to these special operators and introduced boundary conditions in terms of his celebrated dichotomy between limit points and limit circles.

In the 1920s John von Neumann established a general spectral theorem for unbounded self-adjoint operators, which Kunihiko Kodaira used to streamline Weyl's method. Kodaira also generalised Weyl's method to singular ordinary differential equations of even order and obtained a simple formula for the spectral measure.

The same formula had also been obtained independently by E. C. Titchmarsh in 1946 (scientific communication between Japan and the United Kingdom had been interrupted by World War II). Titchmarsh had followed the method of the German mathematician Emil Hilb, who derived the eigenfunction expansions using complex function theory instead of operator theory.

Other methods avoiding the spectral theorem were later developed independently by Levitan, Levinson and Yoshida, who used the fact that the resolvent of the singular differential operator could be approximated by compact resolvents corresponding to Sturm–Liouville problems for proper subintervals. Another method was found by Mark Grigoryevich Krein; his use of direction functionals was subsequently generalised by I. M. Glazman to arbitrary ordinary differential equations of even order. Weyl applied his theory to Carl Friedrich Gauss's hypergeometric differential equation, thus obtaining a far-reaching generalisation of the transform formula of Gustav Ferdinand Mehler (1881) for the Legendre differential equation, rediscovered by the Russian physicist Vladimir Fock in 1943, and usually called the Mehler–Fock transform. The corresponding ordinary differential operator is the radial part of the Laplacian operator on 2-dimensional hyperbolic space.

More generally, the Plancherel theorem forSL(2,R) of Harish Chandra and Gelfand–Naimark can be deduced from Weyl's theory for the hypergeometric equation, as can the theory of spherical functions for the isometry groups of higher dimensional hyperbolic spaces.

Harish Chandra's later development of the Plancherel theorem for general real semisimple Lie groups was strongly influenced by the methods Weyl developed for eigenfunction expansions associated with singular ordinary differential equations. Equally importantly the theory also laid the mathematical foundations for the analysis of the Schrödinger equation and scattering matrix in quantum mechanics.

Solutions of Ordinary Differential Equations

Reduction to Standard Form: Let D be the second order differential operator on (a,b) given by

$$Df(x) = -p(x)f''(x) + r(x)f'(x) + q(x)f(x),$$

where p is a strictly positive continuously differentiable function and q and r are continuous real-valued functions.

For x_0 in (a, b), define the Liouville transformation ψ by

$$\psi(x) = \int_{x_0}^{x} p(t)^{-1/2}\, dt$$

Let U be the unitary operator $(Uf) \cdot\psi\ \psi'^{1/2} = f$ from $L^2\ (a,b)$ onto $L^2(c,d)$, where $c = \psi(a)$ and $d = \psi(b)$. Then

$$UDU^{-1}g = -g'' + Rg' + Qg,$$

where $Q \cdot\psi = q$ and $R \cdot\psi = (p'/2 + r)/p^{1/2}$. The term in g' can be removed using an Euler integrating factor. If S'/S = "$R/2$, then $h = Sg$ satisfies

$$(SUDU^{-1}S^{-1})h = -h'' + Vh,$$

where the potential V is given by $V = Q + S^2/2 + S'/2$. The differential operator can thus always be reduced to one of the form

$$Df = -f'' + qf.$$

Existence Theorem

The following is a version of the classical Picard existence theorem for second order differential equations with values in a Banach space E. Let α, β be arbitrary elements of E, A a bounded operator on E and q a continuous function on $[a,b]$. Then, for $c = a$ or b, the differential equation

$$Df = Af$$

has a unique solution f in $C^2([a,b],E)$ satisfying the initial conditions

$$f(c) = \beta,\ f'(c) = \alpha.$$

In fact a solution of the differential equation with these initial conditions is equivalent to a solution of the integral equation

$$f = h + T f$$

with T the bounded linear map on $C([a,b], E)$ defined by

$$Tf(x) = \int_c^x K(x,y)f(y)dy,$$

where K is the Volterra kernel

$$K(x,t) = (x - t)(q(t) - A)$$

and

$$h(x) = \alpha(x - c) + \beta.$$

Since $||T^k||$ tends to 0, this integral equation has a unique solution given by the Neumann series

$$f = (I - T)^{-1} h = h + T h + T^2 h + T^3 h + \cdots$$

This iterative scheme is often called *Picard iteration* after the French mathematician Charles Emile Picard.

Fundamental Eigenfunctions

If f is twice continuously differentiable (i.e. C^2) on (a, b) satisfying $Df = \lambda f$, then f is called an eigenfunction of L with eigenvalue λ.

- In the case of a compact interval $[a, b]$ and q continuous on $[a, b]$, the existence theorem implies that for $c = a$ or b and every complex number λ there a unique C^2 eigenfunction f_λ on $[a, b]$ with $f_\lambda(c)$ and $f'_\lambda(c)$ prescribed. Moreover, for each x in $[a, b]$, $f_\lambda(x)$ and $f'_\lambda(x)$ are holomorphic functions of λ.

- For an arbitrary interval (a,b) and q continuous on (a, b), the existence theorem implies that for c in (a, b) and every complex number λ there a unique C^2 eigenfunction f_λ on (a, b) with $f_\lambda(c)$ and $f'_\lambda(c)$ prescribed. Moreover, for each x in (a, b), $f_\lambda(x)$ and $f'_\lambda(x)$ are holomorphic functions of λ.

Green's Formula

If f and g are C^2 functions on (a, b), the Wronskian $W(f, g)$ is defined by

$$W(f, g)(x) = f(x)\, g'(x) - f'(x)\, g(x).$$

Green's formula states that for x, y in (a, b)

$$\int_x^y (Df)g - f(Dg)dt = W(f,g)(y) - W(f,g)(x).$$

When q is continuous and f, g C^2 on the compact interval $[a, b]$, this formula also holds for $x = a$ or $y = b$. When f and g are eigenfunctions for the same eigenvalue, then

$$\frac{d}{dx}W(f,g) = 0,$$

so that $W(f, g)$ is independent of x.

Classical Sturm–Liouville Theory

Let $[a, b]$ be a finite closed interval, q a real-valued continuous function on $[a, b]$ and let H_0 be the space of C^2 functions f on $[a, b]$ satisfying the mixed boundary conditions

$$\cos\alpha f(a) - \sin\alpha f'(a) = 0, \qquad \cos\beta f(b) - \sin\beta f'(b) = 0,$$

with inner product

$$(f,g) = \int_a^b f(x)\overline{g(x)}dx.$$

In practise usually one of the two unmixed boundary conditions:

- Dirichlet boundary condition $f(c) = 0$
- Neumann boundary condition $f'(c) = 0$

is imposed at each endpoint $c = a, b$.

The differential operator D given by

$$Df = -f'' + qf$$

acts on H_0. A function f in H_0 is called an eigenfunction of D (for the above choice of boundary values) if $Df = \lambda f$ for some complex number λ, the corresponding eigenvalue. By Green's formula, D is formally self-adjoint on H_0:

$$(Df, g) = (f, Dg) \text{ for } f, g \text{ in } H_0.$$

As a consequence, exactly as for a self-adjoint matrix in finite dimensions,

- the eigenvalues of D are real;
- the eigenspaces for distinct eigenvalues are orthogonal.
- It turns out that the eigenvalues can be described by the maximum-minimum principle of Rayleigh–Ritz. In fact it is easy to see *a priori* that the eigenvalues are bounded below because the operator D is itself *bounded below* on H_0:
- $(Df, f) \geq M(f, f)$ for some finite constant M.

In fact integrating by parts

$$(Df, f) = [-f'\bar{f}]_a^b + \int |f'|^2 + \int q|f|^2.$$

For unmixed boundary conditions, the first term vanishes and the inequality holds with $M = \inf q$. In the mixed case the first term can be estimated using an elementary *Peter-Paul* version of Sobolev's inequality:

"*Given $\Sigma > 0$, there is constant $R > 0$ such that $|f(x)|^2 \leq \Sigma$ $(f', f') + R\ (f, f)$ for all f in $C^1[a, b]$.*"

In fact, since

$$|f(b) - f(x)| \leq (b - a)^{1/2} \cdot \| f' \|_2,$$

only an estimate for $f(b)$ is needed and this follows by replacing $f(x)$ in the above inequality by $(x - a)^n \cdot (b - a)^{-n} f(x)$ for n sufficiently large.

Green's Function (Regular Case)

From the theory of ordinary differential equations, there are unique fundamental eigenfunctions $\varphi_\lambda(x)$, $\chi_\lambda(x)$ such that

- $D\varphi_\lambda = \lambda\varphi_\lambda$, $\varphi_\lambda(a) = \sin\alpha$, $\varphi_\lambda'(a) = \cos\alpha$
- $D\chi_\lambda = \lambda\chi_\lambda$, $\chi_\lambda(b) = \sin\beta$, $\chi_\lambda'(b) = \cos\beta$

which at each point, together with their first derivatives, depend holomorphically on λ. Let

$$\omega(\lambda) = W(\varphi_\lambda, \chi_\lambda),$$

an entire holomorphic function.

This function $\omega(\lambda)$ plays the role of the characteristic polynomial of D. Indeed the uniqueness of the fundamental eigenfunctions implies that its zeros are precisely the eigenvalues of D and that each non-zero eigenspace is one-dimensional. In particular there are at most countably many eigenvalues of D and, if there are infinitely many, they must tend to infinity.

It turns out that the zeros of $\omega(\lambda)$ also have mutilplicity one. If λ is not an eigenvalue of D on H_0, define the Green's function by

$G_\lambda(x,y) = \varphi_\lambda(x)\, \chi_\lambda(y) / \omega(\lambda)$ for $x \geq y$ and $\chi_\lambda(x)\, \varphi_\lambda(y) / \omega(\lambda)$ for $y \geq x$.

This kernel defines an operator on the inner product space $C[a,b]$ via

$$(G_\lambda f)(x) = \int_a^b G_\lambda(x,y) f(y)dy.$$

Since $G_\lambda(x,y)$ is continuous on $[a, b] \times [a, b]$, it defines a Hilbert–Schmidt operator on the Hilbert space completion H of $C[a, b] = H_1$ (or equivalently of the dense subspace H_0), taking values in H_1. This operator carries H_1 into H_0. When λ is real, $G_\lambda(x,y) = G_\lambda(y,x)$ is also real, so defines a self-adjoint operator on H. Moreover

- $G_\lambda (D - \lambda) = I$ on H_0
- G_λ carries H_1 into H_0, and $(D - \lambda)\, G_\lambda = I$ on H_1.

Thus the operator G_λ can be identified with the resolvent $(D - \lambda)^{-1}$.

Spectral Theorem

Theorem: The eigenvalues of D are real of multiplicity one and form an increasing sequence $\lambda_1 < \lambda_2 < \cdots$ tending to infinity.

The corresponding normalised eigenfunctions form an orthonormal basis of H_0.

The kth eigenvalue of D is given by the minimax principle

$$\lambda_k = \max_{\dim G = k-1} \min_{f \perp G} \frac{(Df, f)}{(f, f)}.$$

In particular if $q_1 \leq q_2$, then

$$\lambda_k(D_1) \leq \lambda_k(D_2).$$

In fact let $T = G_\lambda$ for λ large and negative. Then T defines a compact self-adjoint operator on the Hilbert space H. By the spectral theorem for compact self-adjoint operators, H has an orthonormal basis consisting of eigenvectors ψ_n of T with $T\psi_n = \mu_n \psi_n$, where μ_n tends to zero. The range of T contains H0 so is dense. Hence 0 is not an eigenvector of T. The resolvent properties of T imply that ψ_n lies inH_0 and that

$$D \psi_n = (\lambda + 1/\mu_n) \psi_n$$

The minimax principle follows because if

$$\lambda(G) = \min_{f \perp G} \frac{(Df, f)}{(f, f)},$$

then $\lambda(G) = \lambda_k$ for the linear span of the first k -1 eigenfunctions. For any other (k -1)-dimensional subspace G, some f in the linear span of the first k eigenvectors must be orthogonal to G. Hence $\lambda(G) \leq (Df,f)/(f,f) \leq \lambda_k$.

Wronskian as a Fredholm Determinant

For simplicity, suppose that $m \leq q(x) \leq M$ on $[0,\pi]$ with Dirichlet boundary conditions. The minimax principle shows that

$$n^2 + m \leq \lambda_n(D) \leq n^2 + M.$$

It follows that the resolvent $(D - \lambda)^{-1}$ is a trace-class operator whenever λ is not an eigenvalue of D and hence that the Fredholm determinant det I " $\mu(D - \lambda)^{-1}$ is defined. The Dirichlet boundary conditions imply that

$$\omega(\lambda) = \varphi_\lambda(b).$$

Using Picard iteration, Titchmarsh showed that $\varphi_\lambda(b)$, and hence $\omega(\lambda)$, is an entire function of finite order 1/2:

$$\omega(\lambda) = O(\geq^{|\lambda|})$$

At a zero μ of $\omega(\lambda)$, $\varphi_\mu(b) = 0$. Moreover

$$\psi(x) = \partial_\lambda \varphi_\lambda(x)|_{\lambda=\mu}$$

satisfies $(D - \mu)\psi = \varphi_\mu$. Thus

$$\omega(\lambda) = (\lambda - \mu)\psi(b) + O((\lambda - \mu)^2).$$

This implies that

- μ is a simple zero of $\omega(\lambda)$.

For otherwise $\psi(b) = 0$, so that ψ would have to lie in H_0. But then

$$(\varphi_\mu, \varphi_\mu) = ((D - \mu)\psi, \varphi_\mu) = (\psi, (D - \mu)\varphi_\mu) = 0,$$

a contradiction. On the other hand the distribution of the zeros of the entire function $\omega(\lambda)$ is already known from the minimax principle.

By the Hadamard factorisation theorem, it follows that

- $\omega(\lambda) = C\prod(1 - \lambda / \lambda_n)$,

for some non-zero constant C.

Hence

$$\det(I - \mu(D - \lambda)^{-1}) = \prod\left(1 - \frac{\mu}{\lambda_n - \lambda}\right) = \prod\frac{1 - (\lambda + \mu)/\lambda_n}{1 - \lambda / \lambda_n} = \frac{\omega(\lambda + \mu)}{\omega(\lambda)}.$$

In particular if 0 is not an eigenvalue of D

$$\omega(\mu) = \omega(0)\cdot\det(I - \mu D^{-1}).$$

Tools from Abstract Spectral Theory

Functions of Bounded Variation: A function $\rho(x)$ of bounded variation on a closed interval $[a, b]$ is a complex-valued function such that its total variation $V(\rho)$, the supremum of the variations

$$\sum_{r=0}^{k-1} |\rho(x_{r+1}) - \rho(x_r)|$$

over all dissections

$$a = x_0 < x_1 < \ldots < x_k = b$$

is finite. The real and imaginary parts of ρ are real-valued functions of bounded variation. If ρ is real-valued and normalised so that $\rho(a)=0$, it has a canonical decomposition as the difference of two bounded non-decreasing functions:

$$\rho(x) = \rho_+(x) - \rho_-(x),$$

where $\rho_+(x)$ and $\rho_-(x)$ are the total positive and negative variation of ρ over $[a, x]$.

If f is a continuous function on $[a, b]$ its Riemann–Stieltjes integral with respect to ρ

$$\int_a^b f(x)d\rho(x)$$

is defined to be the limit of approximating sums

$$\sum_{r=0}^{k-1} f(x_r)(\rho(x_{r+1})-\rho(x_r))$$

as the mesh of the dissection, given by sup $|x_{r+1} - x_r|$, tends to zero.

This integral satisfies

$$\left|\int_a^b f(x)d\rho(x)\right| \le V(\rho)\cdot\| f \|_\infty$$

and thus defines a bounded linear functional $d\rho$ on $C[a, b]$ with norm $\|d\rho\|=V(\rho)$.

Every bounded linear functional μ on $C[a, b]$ has an absolute value $|\mu|$ defined for non-negative f by

$$|\mu|(f) = \sup_{0\le|g|\le f} |\mu(g)|.$$

The form $|\mu|$ extends linearly to a bounded linear form on $C[a, b]$ with norm $\||\mu|\|$ and satisfies the characterising inequality

$$|\mu(f)| \le |\mu|(|f|)$$

for f in $C[a, b]$. If μ is *real*, i.e. is real-valued on real-valued functions, then

$$\mu = |\mu| - (|\mu| - \mu) \equiv \mu_+ - \mu_-$$

gives a canonical decomposition as a difference of *positive* forms, i.e. forms that are non-negative on non-negative functions. Every positive form μ extends uniquely to the linear span of non-negative bounded lower semicontinuous functions g by the formula

$$\mu(g) = \lim \mu(f_n),$$

where the non-negative continuous functions f_n increase pointwise to g.

The same therefore applies to an arbitrary bounded linear form μ, so that a function ρ of bounded variation may be defined by

$$\rho(x) = \mu(\chi_{[a,x]}),$$

where χ_A denotes the characteristic function of a subset A of $[a, b]$. Thus $\mu = d\rho$ and $||\mu|| = ||d\rho||$. Moreover $\mu_+ = d\rho_+$ and $\mu_- = d\rho_-$. This correspondence between functions of bounded variation and bounded linear forms is a special case of the Riesz representation theorem.

The support of $\mu = d\rho$ is the complement of all points x in $[a,b]$ where ρ is constant on some neighbourhood of x; by definition it is a closed subset A of $[a,b]$. Moreover $\mu((1-\chi_A)f) = 0$, so that $\mu(f) = 0$ if f vanishes on A.

Spectral Measure

Let H be a Hilbert space and T a self-adjoint bounded operator on H with $0 \leq T \leq I$, so that the spectrum $\sigma(T)$ of T is contained in $[0,1]$. If $p(t)$ is a complex polynomial, then by the *spectral mapping theorem*

$$\sigma(p(T)) = p(\sigma(T))$$

and hence

$$||p(T)|| \leq ||p||_\infty,$$

where $||\cdot||_\infty$ denotes the uniform norm on C[0,1]. By the Weierstrass approximation theorem, polynomials are uniformly dense in C[0, 1]. It follows that $f(T)$ can be defined for every f in C[0,1], with

$$\sigma(f(T)) = f(\sigma(T)) \text{ and } ||f(T)|| \leq ||f||_\infty.$$

If $0 \leq g \leq 1$ is a lower semicontinuous function on [0,1], for example the characteristic function $\chi_{[0,\alpha]}$ of a subinterval of [0,1], then g is a pointwise increasing limit of non-negative f_n in C[0,1].

According to Sz.-Nagy, if ξ is a vector in H, then the vectors

$$\eta_n = f_n(T)\ \xi$$

form a Cauchy sequence in H, since, for $n \geq m$,

$$\| \eta_n - \eta_m \|^2 \leq (\eta_n, \xi) - (\eta_m, \xi),$$

and $(\eta_n, \xi) = (f_n(\mathrm{T})\xi, \xi)$ is bounded and increasing, so has a limit.

It follows that $g(T)$ can be defined by

$$g(T)\xi = \lim f_n(T)\xi.$$

If ξ and η are vectors in H, then

$$\mu_{\xi,\eta}(f) = (f(T)\xi, \eta)$$

defines a bounded linear form $\mu_{\xi,\eta}$ on H. By the Riesz representation theorem

$$\mu_{\xi,\eta} = d\rho_{\xi,\eta}$$

for a unique normalised function $\rho_{\xi,\eta}$ of bounded variation on [0,1].

$d\rho_{\xi,\eta}$ (or sometimes slightly incorrectly $\rho_{\xi,\eta}$ itself) is called the spectral measure determined by ξ and η.

The operator $g(T)$ is accordingly uniquely characterised by the equation

$$(g(T)\xi,\eta) = \mu_{\xi,\eta}(g) = \int_0^1 g(\lambda)d\rho_{\xi,\eta}(\lambda).$$

The spectral projection $E(\lambda)$ is defined by

$$E(\lambda) = \chi_{[0,\lambda]}(T),$$

so that

$$\rho_{\xi,\eta}(\lambda) = (E(\lambda)\xi,\eta).$$

It follows that

$$g(T) = \int_0^1 g(\lambda)dE(\lambda),$$

which is understood in the sense that for any vectors ξ and η,

$$(g(T)\xi,\eta) = \int_0^1 g(\lambda)d(E(\lambda)\xi,\eta) = \int_0^1 g(\lambda)d\rho_{\xi,\eta}(\lambda).$$

For a single vector ξ, $\mu_\xi = \mu_{\xi,\xi}$ is a positive form on [0,1] (in other words proportional to a probability measure on [0,1]) and $\rho_\xi = \rho_{\xi,\xi}$ is non-negative and non-decreasing. Polarisation shows that all the forms $\mu_{\xi,\eta}$ can naturally be expressed in terms of such positive forms, since

$$\mu_{\xi,\eta} = \tfrac{1}{4}(\mu_{\xi+\eta} + i\,\mu_{\xi+i\eta} - \mu_{\xi-\eta} - i\,\mu_{\xi-i\eta}).$$

If the vector ξ is such that the linear span of the vectors $(T^n\xi)$ is dense in H, i.e. ξ is a *cyclic vector* for T, then the map U defined by

$$U(f) = f(T)\xi, C[0,1] \to H$$

satisfies

$$(Uf_1, Uf_2) = \int_0^1 f_1(\lambda)\overline{f_2(\lambda)}d\rho_\xi(\lambda).$$

Let $L^2([0,1], d\rho_\xi)$ denote the Hilbert space completion of C[0,1] associated with the possibly degenerate inner product on the right hand side. Thus U extends to a unitary transformation of $L^2([0,1], d\rho_\xi)$ onto *H*. *UTU** is then just multiplication by λ on $L^2([0,1], d\rho_\xi)$; and more generally Uf(T)U* is multiplication by $f(\lambda)$. In this case, the support of $d\rho_\xi$ is exactly $\sigma(T)$, so that

- the self-adjoint operator becomes a multiplication operator on the space of functions on its spectrum with inner product given by the spectral measure.

Weyl–Kodaira Theory

The eigenfunction expansion associated with singular differential operators of the form

$$Df = -(pf')' + qf$$

on an open interval (a, b) requires an initial analysis of the behaviour of the fundamental eigenfunctions near the endpoints a and b to determine possible boundary conditions there.

Unlike the regular Sturm–Liouville case, in some circumstances spectral values of D can have multiplicity 2. In the development outlined below standard assumptions will be imposed on p and q that guarantee that the spectrum of$D$ has multiplicity one everywhere and is bounded below. This includes almost all important applications; modifications required for the more general case will be discussed later.

Having chosen the boundary conditions, as in the classical theory the resolvent of D, $(D + R\ )^{-1}$ for Rl arge and positive, is given by an operator T corresponding to a Green's function constructed from two fundamental eigenfunctions.

In the classical case T was a compact self-adjoint operator; in this *case* T is just a self-adjoint bounded operator with $0 \leq T \leq I$. The abstract theory of spectral measure can therefore be applied to T to give the eigenfunction expansion for D. The central idea in the proof of Weyl and Kodaira can be explained informally as follows. Assume that the spectrum of D lies in $[1,\infty)$ and that $T = D^{-1}$ and let

$$E(\lambda) = \chi_{[\lambda^{-1},1]}(T)$$

be the spectral projection of D corresponding to the interval $[1,\lambda]$. For an arbitrary function f define

$$f(x,\lambda) = (E(\lambda)f)(x).$$

$f(x,\lambda)$ may be regarded as a differentiable map into the space of functions of bounded variation ρ; or equivalently as a differentiable map

$$x \mapsto (d_\lambda f)(x)$$

into the Banach space E of bounded linear functionals $d\rho$ on $C[\alpha,\beta]$ whenever $[\alpha,\beta]$ is a compact subinterval of [1, “).

Weyl's fundamental observation was that $d_\lambda f$ satisfies a second order ordinary differential equation taking values in E:

$$D(d_\lambda f) = \lambda \cdot d_\lambda f.$$

After imposing initial conditions on the first two derivatives at a fixed point c, this equation can be solved explicitly in terms of the two fundamental eigenfunctions and the "initial value" functionals

$$(d_\lambda f)(c) = d_\lambda f(c,\cdot), \quad (d_\lambda f)'(c) = d_\lambda f_x(c,\cdot).$$

This point of view may now be turned on its head: $f(c,\lambda)$ and $f_x(c,\lambda)$ may be written as

$$f(c,\lambda) = (f,\xi_1(\lambda)), \quad f_x(c,\lambda) = (f,\xi_2(\lambda)),$$

where $\xi_1(\lambda)$ and $\xi_2(\lambda)$ are given purely in terms of the fundamental eigenfunctions. The functions of bounded variation

$$\sigma_{ij}(\lambda) = (\xi_i(\lambda),\xi_j(\lambda))$$

determine a spectral measure on the spectrum of D and can be computed explicitly from the behaviour of the fundamental eigenfunctions (the Titchmarsh–Kodaira formula).

Limit Circle and Limit Point for Singular Equations

Let $q(x)$ be a continuous real-valued function on $(0,_\infty)$ and let D be the second order differential operator

$$Df = -f'' + qf$$

on $(0,\infty)$. Fix a point c in $(0,\infty)$ and, for λ complex, let φ_λ, χ_λ be the unique fundamental eigenfunctions of D on $(0,\infty)$ satisfying

$$(D-\lambda)\varphi_\lambda = 0, \quad (D-\lambda)\chi_\lambda = 0$$

together with the initial conditions at c

$$\varphi_\lambda(c) = 1, \varphi_\lambda'(c) = 0, \chi_\lambda(c) = 0, \chi_\lambda'(c) = 1.$$

Then their Wronskian satisfies

$$W(\varphi_\lambda, \chi_\lambda) = \varphi_\lambda \chi_\lambda' - \chi_\lambda \varphi_\lambda' \equiv 1,$$

since it is constant and equal to 1 at c.

$$2\,\mathrm{Im}(\lambda)\int_c^x |\varphi + \mu\chi|^2 = -\mathrm{Im}(\mu).$$

This defines a circle in the complex μ-plane, the interior of which is given by

$$\int_c^x |\varphi + \mu\chi|^2 < -\frac{\mathrm{Im}(\mu)}{2\,\mathrm{Im}(\lambda)}$$

if $x > c$ and by

$$\int_x^c |\varphi + \mu\chi|^2 < -\frac{\mathrm{Im}(\mu)}{2\,\mathrm{Im}(\lambda)}$$

if $x < c$.

Let D_x be the closed disc enclosed by the circle. By definition these closed discs are nested and decrease as x approaches 0 or ∞. So in the limit, the circles tend either to a limit circle or a limit point at each end. In particular if μ is a limit point or a point on the limit circle at 0 or ", then $|\varphi + \mu\chi|^2$is square integrable (L^2) near 0 or ∞. In particular:

- there are always non-zero solutions of $Df = \lambda f$ which are square integrable near 0 or ∞;
- there is exactly one non-zero solution (up to scalar multiples) of $Df = \lambda f$ which is square integrable near 0 or ∞ precisely in the limit point case. On the other hand if $Dg = \lambda' g$ for another value λ', then

  $$h(x) = g(x) - (\lambda' - \lambda)\int_c^x (\varphi_\lambda(x)\chi_\lambda(y) - \chi_\lambda(x)\varphi_\lambda(y))g(y)dy$$

  satisfies $Dh = \lambda h$, so that

$$g(x) = c_1\varphi_\lambda + c_2\chi_\lambda + (\lambda' - \lambda)\int_c^x (\varphi_\lambda(x)\chi_\lambda(y) - \chi_\lambda(x)\varphi_\lambda(y))g(y)dy.$$

Using this to estimate g, it follows that

- the limit point/limit circle behaviour at 0 or ∞ is independent of the choice of λ.

More generally if $Dg = (\lambda - r)\, g$ for some function $r(x)$, then

$$g(x) = c_1\varphi_\lambda + c_2\chi_\lambda - \int_c^x (\varphi_\lambda(x)\chi_\lambda(y) - \chi_\lambda(x)\varphi_\lambda(y))r(y)g(y)dy.$$

From this it follows tha

- if r is continuous at *0*, then $D + r$ is limit point or limit circle at *0* precisely when D is,

  so that in particular

- *if $q(x)$- a/x^2 is continuous at 0, then D is limit point at 0 if and only if $a \geq$ ¾.*

Similarly,

- *if r has a finite limit at “, then D + r is limit point or limit circle at “ precisely when D is,*

so that in particular

- *if q has a finite limit at “, then D is limit point at ∞.*

Many more elaborate criteria to be limit point or limit circle can be found in the mathematical literature.

Green's Function (Singular Case)

Consider the differential operator

$$D_0 f = -(p_0 f')' + q_0 f$$

on $(0,\infty)$ with q_0 positive and continuous on $(0,\infty)$ and p_0 continuously differentiable in $[0,\infty)$, positive in $(0,\infty)$ and $p_0(0)=0$.

Moreover assume that after reduction to standard form D_0 becomes the equivalent operator

$$Df = -f'' + qf$$

on $(0,\infty)$ where q has a finite limit at ∞. Thus

- *D is limit point at ∞.*

At *0*, D may be either limit circle or limit point. In either case there is an eigenfunction ϕ_0 with $D\ \phi_0 = 0$ and ϕ_0 square integrable near 0. In the limit circle case, ϕ_0 determines a boundary condition at 0:

$$W(f,\phi_0)(0) = 0.$$

For λ complex, let ϕ_λ and $\times_\lambda$ satisfy

- $(D - \lambda)\ \phi_\lambda = 0$, $(D - \lambda)\times_\lambda = 0$
- $\times_\lambda$ square integrable near infinity
- ϕ_λ square integrable at 0 if 0 is *limit point*
- ϕ_λ satisfies the boundary condition above if 0 is *limit circle.*

Let

$$\omega(\lambda) = W(\phi_\lambda, \times_\lambda),$$

a constant which vanishes precisely when ϕ_λ and $\times_\lambda$ are proportional, i.e. λ is an eigenvalue of D for these boundary conditions. On the other hand, this cannot occur if Im $\lambda \neq 0$ or if λ is negative.

Indeed if $Df = \lambda f$ with $q_0 - \lambda \geq \delta > 0$, then by Green's formula $(Df,f) = (f,Df)$, since $W(f,f^*)$ is constant. So λ must be real. If f is taken to be real-valued in the D_0 realisation, then for $0 < x < y$

$$[p_0 ff']_x^y = \int_x^y (q_0 - \lambda)|f|^2 + p_0(f')^2.$$

Since $p_0(0) = 0$ and f is integrable near 0, $p_0 ff'$ must vanish at 0. Setting $x = 0$, it follows that $f(y)\, f'(y) > 0$, so that f^2 is increasing, contradicting the square integrability of f near ∞.

Thus, adding a positive scalar to q, it may be assumed that

$\omega(\lambda) \neq 0$ when λ is not in $[1,\infty)$.

If $\omega(\lambda) \neq 0$, the Green's function $G_\lambda(x,y)$ at λ is defined by

$$G_\lambda(x,y) = \Phi_\lambda(x) X_\lambda(y)/\omega(\lambda)(x \leq y),\ X_\lambda(x)\Phi_\lambda(y)/\omega(\lambda)(x \geq y).$$

and is independent of the choice of $_\lambda$ and $\times_\lambda$.

In the examples there will be a third "bad" eigenfunction ψ_λ defined and holomorphic for λ not in $[1, \infty)$ such that ψ_λ satisfies the boundary conditions at neither 0 nor ∞. This means that for λ not in $[1, \infty)$

- $W(\phi_\lambda, \psi_\lambda)$ is nowhere vanishing;
- $W(\times_\lambda, \psi_\lambda)$ is nowhere vanishing.

In this case $\times_\lambda$ is proportional to $\phi_\lambda + m(\lambda)\, \psi_\lambda$, where

- $m(\lambda) = -\, W(\phi_\lambda, \times_\lambda) \,/\, W(\psi_\lambda, \times_\lambda)$.

Let H_1 be the space of square integrable continuous functions on $(0,\infty)$ and let H_0 be

- the space of C^2 functions f on $(0,\infty)$ of compact support if D is limit point at 0
- the space of C^2 functions f on $(0,\infty)$ with $W(f,\phi_0)=0$ at 0 and with $f = 0$ near ∞ if D is limit circle at 0.

Define $T = G_0$ by

$$(Tf)(x) = \int_0^\infty G_0(x, y) f(y) dy.$$

Then $T\,D = I$ on H_0, $D\,T = I$ on H_1 and the operator D is bounded below on H_0:

$$(Df, f) \geq (f, f).$$

Thus T is a self-adjoint bounded operator with $0 \leq T \leq I$.

Formally $T = D^{-1}$.

The corresponding operators G_λ defined for λ not in $[1,\infty)$ can be formally identified with

$$(D - \lambda)^{-1} = T(I - \lambda T)^{-1}$$

and satisfy $G_\lambda\,(D - \lambda) = I$ on H_0, $(D - \lambda)G_\lambda = I$ on H_1.

Chapter 6

Line Coordinates

In geometry, line coordinates are used to specify the position of a line just as point coordinates (or simply coordinates) are used to specify the position of a point.

Lines in the Plane

There are several possible ways to specify the position of a line in the plane. A simple way is by the pair (m, b) where the equation of the line is $y = mx + b$. Here m is the slope and b is the x-intercept. This system specifies coordinates for all lines that are not vertical. However, it is more common and simpler algebraically to use coordinates (l, m) where the equation of the line is $lx + my + 1 = 0$. This system specifies coordinates for all lines except those that pass through the origin. The geometrical interpretations of l and m are the negative reciprocals of the x and y-intercept respectively.

The exclusion of lines passing through the origin can be resolved by using a system of three coordinates (l, m, n) to specify the line in which the equation, $lx + my + n = 0$. Here l and m may not both be 0. In this equation, only the ratios between l, m and n are significant, in other words if the coordinates are multiplied by a non-zero scalar then line represented remains the same. So (l, m, n) is a system of homogeneous coordinates for the line. If points in the plane are represented by homogeneous coordinates (x, y, z), the equation of the line is $lx + my + nz = 0$. In this context, l, m and n may not all be 0. In particular, $(0, 0, 1)$ represents the line $z = 0$, which is the line at infinity in the projective plane. The coordinates $(0, 1, 0)$ and $(1, 0, 0)$ represent the x and y-axes respectively.

Tangential Equations

Just as $f(x, y) = 0$ can represent a curve as a subset of the points in the plane, the equation $\varphi(l, m) = 0$ represents a subset of the lines on the plain. The set of lines on the plane may, in an abstract sense, be thought of as the set of points in a projective plane, the dual of the original plane. The equation $\varphi(l, m) = 0$ then represents a curve in the dual plane. For a curve $f(x, y) = 0$ in the plane, the tangents to the curve form a curve in the dual space called the dual curve.

If $\varphi(l, m) = 0$ is the equation of the dual curve, then it is called the tangential equation, for the original curve. A given equation $\varphi(l, m) = 0$ represents a curve in the original plane determined as the envelope of the lines that satisfy this equation. Similarly, if $\varphi(l, m, n)$ is a homogeneous function then $\varphi(l, m, n) = 0$ represents a curve in the dual space given in homogeneous coordinates, and may be called the homogeneous tangential equation of the enveloped curve. Tangential equations are useful in the study of curves defined as envelopes, just as Cartesian equations are useful in the study of curves defined as loci.

Tangential Equation of a Point

A linear equation in line coordinates has the form $al + bm + c = 0$, where a, b and c are constants. Suppose (l, m) is a line that satisfies this equation. If c is not 0 then $lx + my + 1 = 0$, where $x = a/c$ and $y = b/c$, so every line satisfying the original equation passes though the point (x, y). Conversely, any line through (x, y) satisfies the original equation, so $al + bm + c = 0$ is the equation of set of lines through (x, y). For a given point (x, y), the equation of the set of lines though it is $lx + my + 1 = 0$, so this may be defined as the tangential equation of the point. Similarly, for a point (x, y, z) given in homogeneous coordinates, then the equation of the point in homogeneous tangential coordinates is $(lx, my, nz) = 0$.

Formulas

The intersection of the lines (l_1, m_1) and (l_2, m_2) is the solution to the linear equations

$$l_1x + m_1y + 1 = 0$$

$$l_2x + m_2y + 1 = 0.$$

By Cramer's rule, the solution is

$$x = \frac{m_1 - m_2}{l_1 m_2 - l_2 m_1}, y = -\frac{l_1 - l_2}{l_1 m_2 - l_2 m_1}.$$

The lines (l_1, m_1), (l_2, m_2), and (l_3, m_3) are concurrent when the determinant

$$\begin{vmatrix} l_1 & m_1 & 1 \\ l_2 & m_2 & 1 \\ l_3 & m_3 & 1 \end{vmatrix} = 0.$$

For homogeneous coordinates, the intersection of the lines (l_1, m_1, n_1) and (l_2, m_2, n_2) is

$$(m_1 n_2 - m_2 n_1, l_2 n_1 - l_1 n_2, l_1 m_2 - l_2 m_1).$$

The lines (l_1, m_1, n_1), (l_2, m_2, n_2) and (l_3, m_3, n_3) are concurrent when the determinant

$$\begin{vmatrix} l_1 & m_1 & n_1 \\ l_2 & m_2 & n_2 \\ l_3 & m_3 & n_3 \end{vmatrix} = 0.$$

Dually, the coordinates of the line containing (x_1, y_1, z_1) and (x_2, y_2, z_2) are

$$(y_1 z_2 - y_2 z_1, x_2 z_1 - x_1 z_2, x_1 y_2 - x_2 y_1).$$

Lines in Three-dimensional Space

For two given points in the plane, (x_1, y_1, z_1) and (x_2, y_2, z_2), the three determinants

$$x_1 y_2 - x_2 y_1, x_1 z_2 - x_2 z_1, y_1 z_2 - y_2 z_1$$

determine the line containing them. Similarly, for two points in three-dimensional space (x_1, y_1, z_1, w_1) and (x_2, y_2, z_2, w_2), the line containing them is determined by the six determinants

$$x_1 y_2 - x_2 y_1, x_1 z_2 - x_1 z_2, y_1 z_2 - y_2 z_1, x_1 w_2 - x_2 w_1, y_1 w_2 - y_2 w_1, z_1 w_2 - z_2 w_1.$$

This is the basis for a system of homogeneous line coordinates in three-dimensional space called *Plücker coordinates.* Six numbers in a set of coordinates only represent a line when they satisfy an additional equation. This system maps the space of lines in three-dimensional space to a projective space of dimension five, but with the additional requirement the space of lines is a manifold of dimension four.

More generally, the lines in n-dimensional projective space are determined by a system of $n(n - 1)/2$ homogeneous coordinates that satisfy a set of $(n - 2)(n - 3)/2$ conditions, resulting in a manifold of dimension $2(n - 1)$.

Log-Polar Coordinates

Log-polar coordinates (or logarithmic polar coordinates) is a coordinate system in two dimensions, where a point is identified by two numbers, one for the logarithm of the distance to a certain point, and one for an angle. Log-polar coordinates are closely connected to polar coordinates, which are usually used to describe domains in the plane with some sort of rotational symmetry. In areas like harmonic and complex analysis, the log-polar coordinates are more canonical than polar coordinates.

Definition and Coordinate Transformations

Log-polar coordinates in the plane consist of a pair of real numbers (ρ,θ), where ρ is the logarithm of the distance to a given point (the origin) and θ is the angle between a line of reference (the x-axis) and the line through the origin and the point. The angular coordinate is the same as for polar coordinates, while the radial coordinate is transformed according to the rule

$$r = e^{\rho}.$$

where r is the distance to the origin. The formulas for transformation from Cartesian coordinates to log-polar coordinates are given by

$$\begin{cases} \rho = \log\sqrt{x^2 + y^2}, \\ \theta = \arctan y / x \text{ if } x > 0. \end{cases}$$

and the formulas for transformation from log-polar to Cartesian coordinates are

$$\begin{cases} x = e^{\rho}\cos\theta, \\ y = e^{\rho}\sin\theta. \end{cases}$$

By using complex numbers $(x, y) = x + iy$, the latter transformation can be written as

$$x + iy = e^{\rho + i\theta}$$

i.e. the complex exponential function. From this follows that basic equations in harmonic and complex analysis will have the same simple form as in Cartesian coordinates. This is not the case for polar coordinates.

Some Important Equations in Log-polar Coordinates

Laplace's Equation

Laplace's equation in two dimensions is given by

$$\frac{\partial^2 u}{\partial x^2} + \frac{\partial^2 u}{\partial y^2} = 0$$

in Cartesian coordinates. Writing the same equation in polar coordinates gives the more complicated equation

$$r\frac{\partial}{\partial r}\left(r\frac{\partial u}{\partial r}\right) + \frac{\partial^2 u}{\partial \theta^2} = 0$$

or equivalently

$$\left(r\frac{\partial}{\partial r}\right)^2 u + \frac{\partial^2 u}{\partial \theta^2} = 0$$

However, from the relation $r = e^{\rho}$ it follows that $r\frac{\partial}{\partial r} = \frac{\partial}{\partial \rho}$ so Laplace's equation in log-polar coordinates,

$$\frac{\partial^2 u}{\partial \rho^2} + \frac{\partial^2 u}{\partial \theta^2} = 0$$

has the same simple expression as in Cartesian coordinates. This is true for all coordinate systems where the transformation to Cartesian coordinates is given by a conformal mapping.

Thus, when considering Laplace's equation for a part of the plane with rotational symmetry, e.g. a circular disk, log-polar coordinates is the natural choice.

Dirichlet-to-Neumann Function

The latter coordinate system is for instance suitable for dealing with Dirichlet-to-Neumann functions. If the discrete coordinate system is interpreted as an undirected graph in the unit disc, it can be considered as a model for an electrical network.

To every line segment in the graph is associated a conductance given by a function γ. The electrical network will then serve as a discrete model for the Dirichlet problem in the unit disc, where the Laplace equation takes the form of Kirchhoff's law. On the nodes on the boundary of the circle, an electrical potential (Dirichlet data) is defined, which induces an electrical current (Neumann data) through the boundary nodes. The linear function Λ_γ mapping Dirichlet data to Neumann data is called a Dirichlet-to-Neumann function, and depends on the topology and conductance of the network.

In the case with the continuous disc, it follows that if the conductance is homogeneous, let's say $\gamma = 1$ everywhere, then the Dirichlet-to-Neumann function satisfies the following equation

$$\Lambda_\gamma^2 + \frac{\partial^2}{\partial\theta^2} = 0$$

In order to get a good discrete model of the Dirichlet problem, it would be useful to find a graph in the unit disc whose (discrete) Dirichlet-to-Neumann function has the same property. Even though polar coordinates don't give us any answer, this is exactly what the spiral network given by log-polar coordinates provides us with.

Image Analysis

Already at the end of the 1970s, applications for the discrete spiral coordinate system were given in image analysis. To represent an image in this coordinate system rather than in Cartesian coordinates, gives computational advantages when rotating or zooming in an image. Also, the photo receptors in the retina in the human eye are distributed in a way that has big similarities with the spiral coordinate system. It can also be found in the Mandelbrot fractal.

Log-polar coordinates can also be used to construct fast methods for the Radon transform and its inverse.

Orthogonal Coordinates

In mathematics, orthogonal coordinates are defined as a set of d coordinates $q = (q^1, q^2, ..., q^d)$ in which the coordinate surfaces all meet at right angles (note: superscripts are indices, not exponents). A coordinate surface for a particular coordinate q^k is the curve, surface, or hypersurface on which q_k is a constant. For example, the three-dimensional Cartesian coordinates (x, y, z) is an orthogonal coordinate system, since its coordinate surfaces x = constant, y = constant, and z = constant are planes that meet at right angles to one another, i.e., are perpendicular. Orthogonal coordinates are a special but extremely common case of curvilinear coordinates. While vector operations and physical laws are normally easiest to derive in Cartesian coordinates, non-Cartesian orthogonal coordinates are often used instead for the solution of various problems, especially boundary value problems, such as those arising in field theories of quantum mechanics, fluid flow, electrodynamics and the diffusion of chemical species or heat.

The chief advantage of non-Cartesian coordinates is that they can be chosen to match the symmetry of the problem. For example, the pressure wave due to an explosion far from the ground (or other barriers) depends on 3D space in Cartesian coordinates, however the pressure predominantly moves away from the centre, so that in spherical coordinates the problem becomes very nearly one dimensional (since the pressure wave dominantly depends only on time and the distance from the centre). Another example is (slow) fluid in a straight circular pipe: in Cartesian coordinates, one has to solve a (difficult) two dimensional boundary value problem involving a partial differential equation, but in cylindrical coordinates the problem becomes one dimensional with an ordinary differential equation instead of a partial differential equation.

The reason to prefer orthogonal coordinates instead of general curvilinear coordinates is simplicity: many complications arise when coordinates are not orthogonal. For example, in orthogonal coordinates many problems may be solved by separation of variables. Separation of variables is a mathematical technique that converts a complex d-dimensional problem into d one-dimensional problems that can be solved in terms of known functions. Many equations can be reduced to Laplace's equation or the Helmholtz equation. Laplace's equation is separable in 13 orthogonal coordinate systems, and the Helmholtz equation is separable in 11 orthogonal coordinate systems.

Orthogonal coordinates never have off-diagonal terms in their metric tensor.

In other words, the infinitesimal squared distance ds^2 can always be written as a scaled sum of the squared infinitesimal coordinate displacements

$$ds^2 = \sum_{k=1}^{d} \left(h_k \, dq^k \right)^2$$

where d is the dimension and the scaling functions (or scale factors)

$$h_k(\mathbf{q}) \stackrel{\text{def}}{=} \sqrt{g_{kk}(\mathbf{q})} = |\mathbf{e}_k|$$

equal the square roots of the diagonal components of the metric tensor, or the lengths of the local basis vectors e_k described below. These scaling functions h_i are used to calculate differential operators in the new coordinates, e.g., the gradient, the Laplacian, the divergence and the curl.

A simple method for generating orthogonal coordinates systems in two dimensions is by a conformal mapping of a standard two-dimensional grid of Cartesian coordinates (x, y).

A complex number $z = x + iy$ can be formed from the real coordinates x and y, where i represents the square root of -1. Any holomorphic function $w = f(z)$ with non-zero complex derivative will produce a conformal mapping; if the resulting complex number is written $w = u + iv$, then the curves of constant u and v intersect at right angles, just as the original lines of constant x and y did.

Orthogonal coordinates in three and higher dimensions can be generated from an orthogonal two-dimensional coordinate system, either by projecting it into a new dimension (*cylindrical coordinates*) or by rotating the two-dimensional system about one of its symmetry axes.

However, there are other orthogonal coordinate systems in three dimensions that cannot be obtained by projecting or rotating a two-dimensional system, such as the ellipsoidal coordinates. More general orthogonal coordinates may be obtained by starting with some necessary coordinate surfaces and considering their orthogonal trajectories.

Basis Vectors

Covariant Vasis

In Cartesian coordinates, the basis vectors are fixed (constant). In the more general setting of curvilinear coordinates, a point in space is specified by the coordinates, and at every such point there is bound a set of basis vectors, which generally are not constant: this is the essence of curvilinear coordinates in general and is a very important concept. What distinguishes orthogonal coordinates is that, though the basis vectors vary, they are always orthogonal with respect to each other. In other words, $\mathbf{e}_i \cdot \mathbf{e}_j = 0$ if $i \neq j$. These basis vectors are by definition the tangent vectors of the curves obtained by varying one coordinate, keeping the others fixed:

$$\mathbf{e}_i = \frac{\partial \mathbf{r}}{\partial q^i}$$

where r is some point and q^i is the coordinate for which the basis vector is extracted. In other words, a curve is obtained by fixing all but one coordinate; the unfixed coordinate is varied as in a parametric curve, and the derivative of the curve with respect to the parametre (the varying coordinate) is the basis vector for that coordinate.

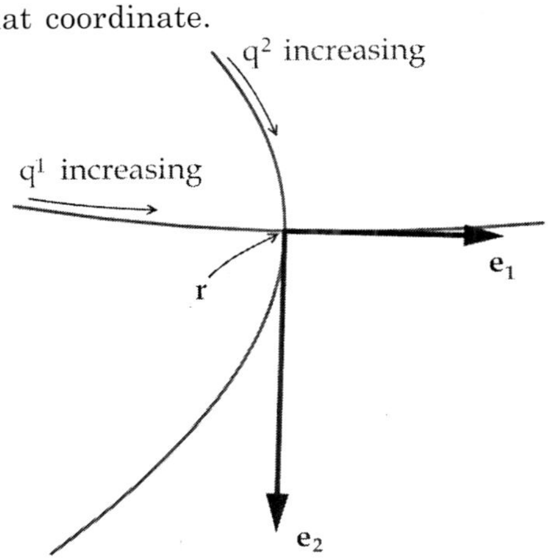

Figure: *Visualization of 2D orthogonal coordinates. Curves obtained by holding all but one coordinate constant are shown, along with basis vectors. Note that the basis vectors aren't of equal length: they need not be, they only need to be orthogonal.*

Note that the vectors are not necessarily of equal length. The normalised basis vectors are notated with a hat and obtained by dividing by the length:

$$\hat{\mathbf{e}}_i = \frac{\mathbf{e}_i}{|\mathbf{e}_i|}$$

A vector field may be specified by its components with respect to the basis vectors or the normalised basis vectors, one must be sure which case is dealt. Components in the normalised basis are most common in applications for clarity of the quantities (for example, one may want to deal with tangential velocity instead of tangential velocity times a scale factor); in derivations the normalised basis is less common since it is more complicated.

The useful functions known as scale factors (sometimes called Lamé coefficients, this should be avoided since some more well known coefficients in linear elasticity carry the same name) of the coordinates are simply the lengths of the basis vectors.

Contravariant Basis

The basis vectors shown above are covariant basis vectors (because they "co-vary" with vectors). In the case of orthogonal coordinates, the contravariant basis vectors are easy to find since they will be in the same direction as the covariant vectors but reciprocal length (for this reason, the two sets of basis vectors are said to be reciprocal with respect to each other):

$$\mathbf{e}^i = \frac{\hat{\mathbf{e}}_i}{h_i} = \frac{\mathbf{e}_i}{h_i^2}$$

this follows from the fact that, by definition, $\mathbf{e}_i \cdot \mathbf{e}_j = \delta_i^j$, using the Kronecker delta. Note that:

$$\hat{\mathbf{e}}_i = \frac{\mathbf{e}_i}{h_i} = h_i \mathbf{e}^i = \hat{\mathbf{e}}^i$$

We now face three different basis sets commonly used to describe vectors in orthogonal coordinates: the covariant basis e_i, the contravariant basis e^i, and the normalised basis κ^i. While a vector is an *objective quantity*, meaning its identity is independent of any coordinate system, the components of a vector depend on what basis the vector is represented in.

To avoid confusion, the components of the vector x with respect to the e_i basis are represented as x^i, while the components with respect to the e^i basis are represented as x_i:

$$\mathbf{x} = \sum x^i \mathbf{e}_i = \sum x_i \mathbf{e}^i$$

The position of the indices represent how the components are calculated (upper indices should not be confused with exponentiation). Note that the summation symbols Σ (capital Sigma) and the summation range, indicating summation over all basis vectors (i = 1, 2, ..., d), are often omitted. The components are related simply by:

$$h_i^2 x^i = x_i$$

There is no distinguishing widespread notation in use for vector components with respect to the normalised basis; in this article we'll use subscripts for vector components and note that the components are calculated in the normalised basis.

Vector Algebra

Vector addition and negation are done component-wise just as in Cartesian coordinates with no complication. Extra considerations may be necessary for other vector operations. Note however, that all of these operations assume that two vectors in a vector field are bound to the same point (in other words, the tails of vectors coincide). Since basis vectors generally vary in orthogonal coordinates, if two vectors are added whose components are calculated at different points in space, the different basis vectors require consideration.

Dot Product

The dot product in Cartesian coordinates (Euclidean space with an orthonormal basis set) is simply the sum of the products of components. In orthogonal coordinates, the dot product of two vectors x and y takes this familiar form when the components of the vectors are calculated in the normalised basis:

$$\mathbf{x} \cdot \mathbf{y} = \sum x_i \hat{\mathbf{e}}_i \cdot \sum y_i \hat{\mathbf{e}}_i = \sum x_i y_i$$

This is an immediate consequence of the fact that the normalised basis at some point can form a Cartesian coordinate system: the basis set is orthonormal.

For components in the covariant or contraviant bases,

$$\mathbf{x}\cdot\mathbf{y} = \sum h_i^2 x^i y^i = \sum \frac{x_i y_i}{h_i^2} = \sum x^i y_i = \sum x_i y^i$$

This can be readily derived by writing out the vectors in component form, normalising the basis vectors, and taking the dot product. For example, in 2D:

$$\mathbf{x}\cdot\mathbf{y} \quad = \left(x^1\mathbf{e}_1 + x^2\mathbf{e}_2\right)\cdot\left(y_1\mathbf{e}^1 + y_2\mathbf{e}^2\right)$$

$$[10pt] \quad = \left(x^1 h_1 \hat{\mathbf{e}}_1 + x^2 h_2 \hat{\mathbf{e}}_2\right)\cdot\left(y_1 \frac{\hat{\mathbf{e}}^1}{h_1} + y_2 \frac{\hat{\mathbf{e}}^2}{h_2}\right) = x^1 y_1 + x^2 y_2$$

where the fact that the normalised covariant and contravariant bases are equal has been used.

Cross Product

The cross product in 3D Cartesian coordinates is:

$$\mathbf{x}\times\mathbf{y} = (x_2y_3 - x_3y_2)\hat{\mathbf{e}}_1 + (x_3y_1 - x_1y_3)\hat{\mathbf{e}}_2 + (x_1y_2 - x_2y_1)\hat{\mathbf{e}}_3$$

The above formula then remains valid in orthogonal coordinates if the components are calculated in the normalised basis. To construct the cross product in orthogonal coordinates with covariant or contravariant bases we again must simply normalise the basis vectors, for example:

$$\mathbf{x}\times\mathbf{y} = \sum x^i\mathbf{e}_i \times \sum y^i\mathbf{e}_i = \sum x^i h_i \hat{\mathbf{e}}_i \times \sum y^i h_i \hat{\mathbf{e}}_i$$

which, written expanded out,

$$\mathbf{x}\times\mathbf{y} = (x^2y^3 - x^3y^2)\frac{h_2h_3}{h_1}\mathbf{e}_1 + (x^3y^1 - x^1y^3)\frac{h_1h_3}{h_2}\mathbf{e}_2 + (x^1y^2 - x^2y^1)\frac{h_1h_2}{h_3}\mathbf{e}_3$$

Terse notation for the cross product, which simplifies generalisation to non-orthogonal coordinates and higher dimensions, is possible with the Levi-Civita tensor, which will have components other than zeros and ones if the scale factors are not all equal to one.

Vector Calculus

Differentiation

Looking at an infinitesimal displacement from some point, it's apparent that

$$d\mathbf{r} = \sum \frac{\partial \mathbf{r}}{\partial q^i} dq^i = \sum \mathbf{e}_i dq^i$$

By definition, the gradient of a function must satisfy (this definition remains true if f is any tensor)

$$df = \nabla f \cdot d\mathbf{r} \quad \Rightarrow \quad df = \nabla f \cdot \sum \mathbf{e}_i dq^i$$

It follows then that del operator must be:

$$\nabla = \sum \mathbf{e}^i \frac{\partial}{\partial q^i}$$

and this happens to remain true in general curvilinear coordinates. Quantities like the gradient and Laplacian follow through proper application of this operator.

Differential Operators in Three Dimensions

Since these operations are common in application, all vector components in this section are presented with respect to the normalised basis.

The gradient of a scalar equals

$$\nabla\phi = \frac{\hat{\mathbf{e}}_1}{h_1}\frac{\partial\phi}{\partial q^1} + \frac{\hat{\mathbf{e}}_2}{h_2}\frac{\partial\phi}{\partial q^2} + \frac{\hat{\mathbf{e}}_3}{h_3}\frac{\partial\phi}{\partial q^3}.$$

The Laplacian of a scalar equals

$$\nabla^2\phi = \frac{1}{h_1h_2h_3}\left[\frac{\partial}{\partial q^1}\left(\frac{h_2h_3}{h_1}\frac{\partial\phi}{\partial q^1}\right) + \frac{\partial}{\partial q^2}\left(\frac{h_3h_1}{h_2}\frac{\partial\phi}{\partial q^2}\right) + \frac{\partial}{\partial q^3}\left(\frac{h_1h_2}{h_3}\frac{\partial\phi}{\partial q^3}\right)\right].$$

The divergence of a vector equals

$$\nabla\cdot\mathbf{F} = \frac{1}{h_1h_2h_3}\left[\frac{\partial}{\partial q^1}(F_1h_2h_3) + \frac{\partial}{\partial q^2}(F_2h_3h_1) + \frac{\partial}{\partial q^3}(F_3h_1h_2)\right].$$

The curl equals

$$\begin{aligned}\nabla\times\mathbf{F} &= \frac{\hat{\mathbf{e}}_1}{h_2h_3}\left[\frac{\partial}{\partial q^2}(h_3F_3) - \frac{\partial}{\partial q^3}(h_2F_2)\right]\\ &+ \frac{\hat{\mathbf{e}}_2}{h_3h_1}\left[\frac{\partial}{\partial q^3}(h_1F_1) - \frac{\partial}{\partial q^1}(h_3F_3)\right]\\ &+ \frac{\hat{\mathbf{e}}_3}{h_1h_2}\left[\frac{\partial}{\partial q^1}(h_2F_2) - \frac{\partial}{\partial q^2}(h_1F_1)\right]\end{aligned}$$

$$[10pt] \quad = \frac{1}{h_1 h_2 h_3} \begin{vmatrix} h_1 \hat{\mathbf{e}}_1 & h_2 \hat{\mathbf{e}}_2 & h_3 \hat{\mathbf{e}}_3 \\ \frac{\partial}{\partial q^1} & \frac{\partial}{\partial q^2} & \frac{\partial}{\partial q^3} \\ h_1 F_1 & h_2 F_2 & h_3 F_3 \end{vmatrix}.$$

Parabolic Cylindrical Coordinates

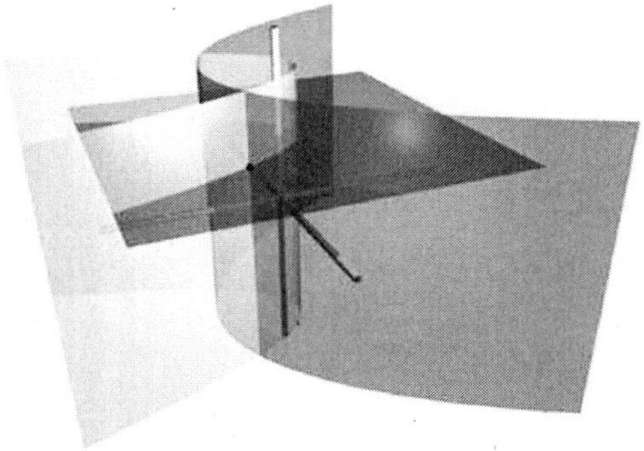

Figure: *Coordinate surfaces of parabolic cylindrical coordinates. The red parabolic cylinder corresponds to σ=2, whereas the yellow parabolic cylinder corresponds to τ=1. The blue plane corresponds to z=2. These surfaces intersect at the point P (shown as a black sphere), which has Cartesian coordinates roughly (2, -1.5, 2).*

In mathematics, parabolic cylindrical coordinates are a three-dimensional orthogonal coordinate system that results from projecting the two-dimensional parabolic coordinate system in the perpendicular z-direction.

Hence, the coordinate surfaces are confocal parabolic cylinders. Parabolic cylindrical coordinates have found many applications, e.g., the potential theory of edges.

Basic Definition

The parabolic cylindrical coordinates (σ, τ, z) are defined in terms of the Cartesian coordinates *(x,y,z)* by:

$$x = \sigma\tau$$

$$y = \frac{1}{2}\left(\tau^2 - \sigma^2\right)$$

$$z = z$$

The surfaces of constant σ form confocal parabolic cylinders

$$2y = \frac{x^2}{\sigma^2} - \sigma^2$$

that open towards $+y$, whereas the surfaces of constant τ form confocal parabolic cylinders

$$2y = -\frac{x^2}{\tau^2} + \tau^2$$

that open in the opposite direction, i.e., towards $-y$. The foci of all these parabolic cylinders are located along the line defined by $x = y = 0$. The radius r has a simple formula as well

$$r = \sqrt{x^2 + y^2} = \frac{1}{2}\left(\sigma^2 + \tau^2\right)$$

that proves useful in solving the Hamilton-Jacobi equation in parabolic coordinates for the inverse-square central force problem of mechanics.

Scale Factors

The scale factors for the parabolic cylindrical coordinates σ and τ are:

$$h_\sigma = h_\tau = \sqrt{\sigma^2 + \tau^2}$$

$$h_z = 1$$

The infinitesimal element of volume is

$$dV = h_\sigma h_\tau h_z = \left(\sigma^2 + \tau^2\right) d\sigma d\tau dz$$

and the Laplacian equals

$$\nabla^2 \Phi = \frac{1}{\sigma^2 + \tau^2}\left(\frac{\partial^2 \Phi}{\partial \sigma^2} + \frac{\partial^2 \Phi}{\partial \tau^2}\right) + \frac{\partial^2 \Phi}{\partial z^2}$$

Other differential operators such as $\nabla \cdot \mathbf{F}$ and $\nabla \times \mathbf{F}$ can be expressed in the coordinates (σ, τ) by substituting the scale factors into the general formulae found in orthogonal coordinates.

Parabolic Cylinder Harmonics

Since all of the surfaces of constant σ, τ and z are conicoid, Laplace's equation is separable in parabolic cylindrical coordinates. Using the technique of the separation of variables, a separated solution to Laplace's equation may be written:

$$V = S(\sigma)T(\tau)Z(z)$$

and Laplace's equation, divided by V, is written:

$$\frac{1}{\sigma^2+\tau^2}\left[\frac{\ddot{S}}{S}+\frac{\ddot{T}}{T}\right]+\frac{\ddot{Z}}{Z}=0$$

Since the Z equation is separate from the rest, we may write

$$\frac{\ddot{Z}}{Z}=-m^2$$

where m is constant. $Z(z)$ has the solution:

$$Z_m(z) = A_1 e^{imz} + A_2 e^{-imz}$$

Substituting $-m^2$ for $\ddot{Z}/Z$, Laplace's equation may now be written:

$$\left[\frac{\ddot{S}}{S}+\frac{\ddot{T}}{T}\right]=m^2(\sigma^2+\tau^2)$$

We may now separate the S and T functions and introduce another constant n^2 to obtain:

$$\ddot{S}-(m^2\sigma^2+n^2)S=0$$

$$\ddot{T}-(m^2\tau^2-n^2)T=0$$

The solutions to these equations are the parabolic cylinder functions

$$S_{mn}(\sigma) = A_3 y_1(n^2/2m, \sigma\sqrt{2m}) + A_4 y_2(n^2/2m, \sigma\sqrt{2m})$$

$$T_{mn}(\tau) = A_5 y_1(n^2/2m, i\tau\sqrt{2m}) + A_6 y_2(n^2/2m, i\tau\sqrt{2m})$$

The parabolic cylinder harmonics for (m,n) are now the product of the solutions. The combination will reduce the number of constants and the general solution to Laplace's equation may be written:

$$V(\sigma,\tau,z) = \sum_{m,n} A_{mn} S_{mn} T_{mn} Z_m$$

Applications

The classic applications of parabolic cylindrical coordinates are in solving partial differential equations, e.g., Laplace's equation or the Helmholtz equation, for which such coordinates allow a separation of variables. A typical example would be the electric field surrounding a flat semi-infinite conducting plate.

Two-centre Bipolar Coordinates

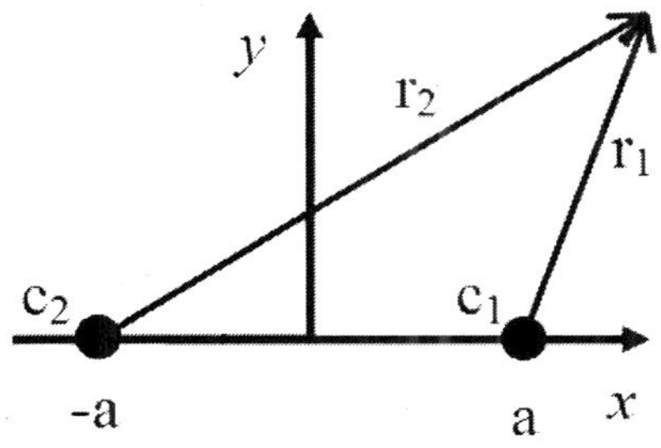

Figure: *Two-centre bipolar coordinates.*

In mathematics, two-centre bipolar coordinates is a coordinate system, based on two coordinates which give distances from two fixed centres, C_1 and C_2 This system is very useful in some scientific applications It should not be confused with so-called bipolar coordinates.

Cartesian coordinates

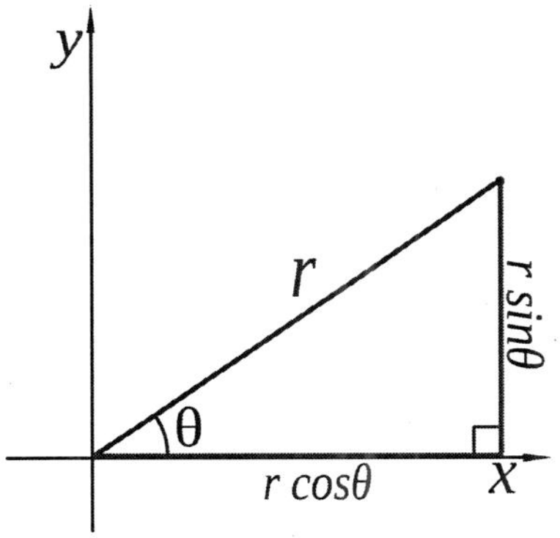

Figure: *Cartesian coordinates and polar coordinates.*

Transformation to Cartesian coordinates (x, y) from two-centre bipolar coordinates (r_1, r_2)

$$x = \frac{r_1^2 - r_2^2}{4a}$$

$$y = \pm \frac{1}{4a}\sqrt{16a^2 r_1^2 - (r_1^2 - r_2^2 + 4a^2)^2}$$

where the centres of this coordinate system are at (+a, 0) and (-a, 0).

Polar Coordinates

To polar coordinates from two-centre bipolar coordinates

$$r = \sqrt{\frac{r_1^2 + r_2^2 - 2a^2}{2}}$$

$$\theta = \arctan\left[\frac{\sqrt{8a^2(r_1^2 + r_2^2 - 2a^2) - (r_1^2 - r_2^2)^2}}{r_1^2 - r_2^2}\right]$$

Where $2a$ is the distance between the poles (coordinate system centres).

Darboux's Theorem

Darboux's theorem is a theorem in the mathematical field of differential geometry and more specifically differential forms, partially generalising the Frobenius integration theorem. It is a foundational result in several fields, the chief among them being symplectic geometry. The theorem is named after Jean Gaston Darboux who established it as the solution of the Pfaff problem. One of the many consequences of the theorem is that any two symplectic manifolds of the same dimension are locally symplectomorphic to one another. That is, every $2n$-dimensional symplectic manifold can be made to look locally like the linear symplectic space C^n with its canonical symplectic form. There is also an analogous consequence of the theorem as applied to contact geometry.

Statement and First Consequences

The precise statement is as follows. Suppose that θ is a differential 1-form on an n dimensional manifold, such that dθ has constant rank p. If

$$\theta \wedge (d\theta)^p = 0 \text{ everywhere,}$$

then there is a local system of coordinates $x_1, \ldots, x_{n-p}$, y_1, ..., y_p in which

$$\theta = x_1 dy_1 + \ldots + x_p dy_p .$$

If, on the other hand,

$$\theta \wedge (d\theta)^p \neq 0 \text{ everywhere,}$$

then there is a local system of coordinates $x_1, \ldots, x_{n-p}, y_1, \ldots, y_p$ in which

$$\theta = x_1 dy_1 + \ldots + x_p dy_p + dx_{p+1} .$$

In particular, suppose that ω is a symplectic 2-form on an $n = 2m$ dimensional manifold M. In a neighbourhood of each point p of M, by the Poincaré lemma, there is a 1-form θ with $d\theta = \omega$. Moreover, θ satisfies the first set of hypotheses in Darboux's theorem, and so locally there is a coordinate chart U near p in which

$$\theta = x_1 dy_1 + \ldots + x_m dy_m .$$

Taking an exterior derivative now shows

$$\omega = d\theta = dx_1 \wedge dy_1 + \ldots + dx_m \wedge dy_m .$$

The chart U is said to be a Darboux chart around p. The manifold M can be covered by such charts.

To state this differently, identify R^{2m} with C^m by letting $z_j = x_j + i\, y_j$. If $\varphi : U \to C^n$ is a Darboux chart, then ω is the pullback of the standard symplectic form ω_0 on C^n:

$$\omega = \phi^* \omega_0 .$$

Comparison with Riemannian Geometry

This result implies that there are no local invariants in symplectic geometry: a Darboux basis can always be taken, valid near any given point.

This is in marked contrast to the situation in Riemannian geometry where the curvature is a local invariant, an obstruction to the metric being locally a sum of squares of coordinate differentials.

The difference is that Darboux's theorem states that ω can be made to take the standard form in an *entire neighbourhood* around p. In Riemannian geometry, the metric can always be made to take the standard form *at* any given point, but not always in a neighbourhood around that point.

Isothermal Coordinates

In mathematics, specifically in differential geometry, isothermal coordinates on a Riemannian manifold are local coordinates where the metric is conformal to the Euclidean metric. This means that in isothermal coordinates, the Riemannian metric locally has the form

$$g = e^{\varphi}(dx_1^2 + \cdots + dx_n^2),$$

where φ is a smooth function.

Isothermal coordinates on surfaces were first introduced by Gauss. Korn and Lichtenstein proved that isothermal coordinates exist around any point on a two dimensional Riemannian manifold. On higher dimensional Riemannian manifolds a necessary and sufficient condition for their local existence is the vanishing of the Weyl tensor and of the Cotton tensor.

Isothermal Coordinates on Surfaces

The first result on the existence of isothermal coordinates was due to Korn (1916) and Lichtenstein (1916). Later accounts were given by Morrey (1938), Ahlfors (1955), Bers (1952) and Chern (1955). A particularly simple account using the Hodge star operator is given in DeTurck & Kazdan (1981).

Beltrami Equation

The existence of isothermal coordinates can be proved by applying known existence theorems for the Beltrami equation, which rely on L^p estimates for singular integral operators of Calderon and Zygmund. A simpler approach to the Beltrami equation has been given more recently by the late Adrien Douady.

If the Riemannian metric is given locally as

$$ds^2 = Edx^2 + 2Fdxdy + Gdy^2,$$

then in the complex coordinate $z = x + iy$, it takes the form

$$ds^2 = \lambda\, |dz + \mu d\bar{z}|^2,$$

where λ and μ are smooth with $\lambda > 0$ and $|\mu| < 1$. In fact

$$\lambda = \frac{1}{4}(E + G + 2\sqrt{EG - F^2}),\ \mu = (E - G + 2iF)/\lambda.$$

In isothermal coordinates (u, v) the metric should take the form

$$ds^2 = \rho(du^2 + dv^2)$$

with ρ > 0 smooth. The complex coordinate $w = u + i\, v$ satisfies

$$\rho |\, dw\, |^2 = \rho\, |\, w_z\, |^2 |dz + \frac{w_{\bar{z}}}{w_z} d\bar{z}\, |^2,$$

so that the coordinates (u, v) will be isothermal if the Beltrami equation

$$\frac{\partial w}{\partial \bar{z}} = \mu \frac{\partial w}{\partial z}$$

has a diffeomorphic solution. Such a solution has been proved to exist in any neighbourhood where $||\mu||_{\infty} < 1$.

Gaussian Curvature

In the isothermal coordinates (u, v), the Gaussian curvature takes the simpler form

$$K = -\frac{1}{2} e^{-\varphi} \left(\frac{\partial^2 \varphi}{\partial u^2} + \frac{\partial^2 \varphi}{\partial v^2} \right),$$

where $\rho = e^{\varphi}$.

Normal Coordinates

In differential geometry, normal coordinates at a point p in a differentiable manifold equipped with a symmetric affine connection are a local coordinate system in a neighbourhood of p obtained by applying the exponential map to the tangent space at p. In a normal coordinate system, the Christoffel symbols of the connection vanish at the point p, thus often simplifying local calculations. In normal coordinates associated to the Levi-Civita connection of a Riemannian manifold, one can additionally arrange that the metric tensor is the Kronecker delta at the point p, and that the first partial derivatives of the metric at p vanish.

A basic result of differential geometry states that normal coordinates at a point always exist on a manifold with a symmetric affine connection. In such coordinates the covariant derivative reduces to a partial derivative (at p only), and the geodesics through p are locally linear functions of t (the affine parametre). This idea was implemented in a fundamental way by Albert Einstein in the general theory of relativity: the equivalence principle uses normal coordinates via inertial frames. Normal coordinates always exist for the Levi-Civita connection of a Riemannian or Pseudo-Riemannian manifold. By contrast, there is no way to define normal coordinates for Finsler manifolds (Busemann 1955).

Geodesic Normal Coordinates

Geodesic normal coordinates are local coordinates on a manifold with an affine connection afforded by the exponential map

$$\exp_p : T_pM \supset V \to M$$

and an isomorphism

$$E : \mathbb{R}^n \to T_pM$$

given by any basis of the tangent space at the fixed basepoint $p - M$. If the additional structure of a Riemannian metric is imposed, then the basis defined by E may be required in addition to be orthonormal, and the resulting coordinate system is then known as a Riemannian normal coordinate system.

Normal coordinates exist on a normal neighbourhood of a point p in M.

A normal neighbourhood U is a subset of M such that there is a proper neighbourhood V of the origin in the tangent space T_pM and $\exp_p$ acts as a diffeomorphism between U and V. Now let U be a normal neighbourhood of p in M then the chart is given by:

$$\varphi := E^{-1} \circ \exp_p^{-1} : U \to \mathbb{R}^n$$

The isomorphism E can be any isomorphism between both vectorspaces, so there are as many charts as different orthonormal bases exist in the domain of E.

Polar Coordinates

On a Riemannian manifold, a normal coordinate system at p facilitates the introduction of a system of spherical coordinates, known as polar coordinates. These are the coordinates on M obtained by introducing the standard spherical coordinate system on the Euclidean space T_pM. That is, one introduces on T_pM the standard spherical coordinate system (r, φ) where $r \geq 0$ is the radial parametre and $\varphi = (\varphi_1,\ldots,\varphi_{n-1})$ is a parameterization of the $(n-1)$-sphere. Composition of (r, φ) with the inverse of the exponential map at p is a polar coordinate system. Polar coordinates provide a number of fundamental tools in Riemannian geometry. The radial coordinate is the most significant: geometrically it represents the geodesic distance to p of nearby points. Gauss's lemma asserts that the gradient of r is simply the partial derivative $\partial / \partial r$. That is,

$$\langle df, dr \rangle = \frac{\partial f}{\partial r}$$

for any smooth function f. As a result, the metric in polar coordinates assumes a block diagonal form

$$g = \begin{bmatrix} 1 & 0 & \cdots 0 \\ 0 & & \\ \vdots & & g_{\phi\phi}(r, \phi) \\ 0 & & \end{bmatrix}.$$

Chapter 7

Vector Field Reconstruction

Vector field reconstruction is a method of creating a vector field from experimental or computer generated data, usually with the goal of finding a differential equation model of the system.

A differential equation model is one that describes the value of dependent variables as they evolve in time or space by giving equations involving those variables and their derivatives with respect to some independent variables, usually time and/or space. An ordinary differential equation is one in which the system's dependent variables are functions of only one independent variable. Many physical, chemical, biological and electrical systems are well described by ordinary differential equations. Frequently we assume a system is governed by differential equations, but we do not have exact knowledge of the influence of various factors on the state of the system.

For instance, we may have an electrical circuit that in theory is described by a system of ordinary differential equations, but due to the tolerance of resistors, variations of the supply voltage or interference from outside influences we do not know the exact parameters of the system. For some systems, especially those that support chaos, a small change in parameter values can cause a large change in the behaviour of the system, so an accurate model is extremely important. Therefore, it may be necessary to construct more exact differential equations by building them up based on the actual system performance rather than a theoretical model. Ideally, one would measure all the dynamical variables involved over an extended period of time, using many different initial conditions, then build or fine tune a differential equation model based on these measurements.

In some cases we may not even know enough about the processes involved in a system to even formulate a model. In other cases, we may have access to only one dynamical variable for our measurements, i.e., we have a scalar time series. If we only have a scalar time series, we need to use the method of time delay embedding or derivative coordinates to get a large enough set of dynamical variables to describe the system.

In a nutshell, once we have a set of measurements of the system state over some period of time, we find the derivatives of these measurements, which gives us a local vector field, then determine a global vector field consistent with this local field. This is usually done by a least squares fit to the derivative data.

Formulation

In the best possible case, one has data streams of measurements of all the system variables, equally spaced in time, say

$$s_1(t),\ s_2(t),\ \ldots\ ,\ s_k(t)$$

for

$$t = t_1,\ t_2, \ldots,\ t_n,$$

beginning at several different initial conditions. Then the task of finding a vector field, and thus a differential equation model consists of fitting functions, for instance, a cubic spline, to the data to obtain a set of continuous time functions

$$x_1(t),\ x_2(t),\ \ldots\ ,\ x_k(t),$$

computing time derivatives dx_1/dt, $dx_2/dt,\ldots,dx_k/dt$ of the functions, then making a least squares fit using some sort of orthogonal basis functions (orthogonal polynomials, radial basis functions, etc.) to each component of the tangent vectors to find a global vector field. A differential equation then can be read off the global vector field.

There are various methods of creating the basis functions for the least squares fit. The most common method is the Gram–Schmidt process. Which creates a set of orthogonal basis vectors, which can then easily be normalized. This method begins by first selecting any standard basis $\beta=\{v_1,\ v_2,\ldots,v_n\}$.

Next, set the first vector $v_1=u_1$. Then, we set $u_2=v_2$-$\text{proj}_{u1}v_2$. This process is repeated to for k vectors, with the final vector being $u_k= v_k$-$\Sigma_{(j=1)}{}^{(k-1)}\text{proj}_{uk}v_k$. This then creates a set of orthogonal standard basis vectors.

The reason for using a standard orthogonal basis rather than a standard basis arises from the creation of the least squares fitting done next. Creating a least-squares fit begins by assuming some function, in the case of the reconstruction an nth degree polynomial, and fitting the curve to the data using constants. The accuracy of the fit can be increased by increasing the degree of the polynomial being used to fit the data. If a set of non-orthogonal standard basis functions was used, it becomes necessary to recalculate the constant coefficients of the function describing the fit. However, by using the orthogonal set of basis functions, it is not necessary to recalculate the constant coefficients.

Applications

Vector field reconstruction has several applications, and many different approaches. Some mathematicians have not only used radial basis functions and polynomials to reconstruct a vector field, but they have used Lyapunov exponents and singular value decomposition. Gouesbet and Letellier used a multivariate polynomial approximation and least squares to reconstruct their vector field. This method was applied to the Rössler system, and the Lorenz system, as well as thermal lens oscillations.

The Rossler system, Lorenz system and Thermal lens oscillation follows the differential equations in standard system as

$$X'=Y, \ Y'=Z \text{ and } Z'=F(X,Y,Z)$$

where $F(X,Y,Z)$ is known as the standard function.

Implementation Issues

In some situation the model is not very efficient and difficulties can arise if the model has a large number of coefficients and demonstrates a divergent solution. For example, nonautonomous differential equations give the previously described results. In this case the modification of the standard approach in application gives a better way of further development of global vector reconstruction.

Usually the system being modelled in this way is a chaotic dynamical system, because chaotic systems explore a large part of the phase space and the estimate of the global dynamics based on the local dynamics will be better than with a system exploring only a small part of the space.

Frequently, one has only a single scalar time series measurement from a system known to have more than one degree of freedom. The time series may not even be from a system variable, but may be instead of a function of all the variables, such as temperature in a stirred tank reactor using several chemical species. In this case, one must use the technique of delay coordinate embedding, where a state vector consisting of the data at time t and several delayed versions of the data is constructed.

Vector-valued Function

A vector-valued function also referred to as a vector function is a mathematical function of one or more variables whose range is a set of multidimensional vectors or infinite-dimensional vectors. The input of a vector-valued function could be a scalar or a vector. The dimension of the domain is not defined by the dimension of the range.

Example

A common example of a vector valued function is one that depends on a single real number parameter t, often representing time, producing a vector $v(t)$ as the result. In terms of the standard unit vectors i, j, k of Cartesian 3-space, these specific type of vector-valued functions are given by expressions such as

- $r(t) = f(t)i + g(t)j$ or
- $r(t) = f(t)i + g(t)j + h(t)k$

where $f(t)$, $g(t)$ and $h(t)$ are the coordinate functions of the parameter t. The vector $r(t)$ has its tail at the origin and its head at the coordinates evaluated by the function.

The vector shown in the graph to the right is the evaluation of the function near t=19.5 (between 6π and 6.5π; i.e., somewhat more than 3 rotations). The spiral is the path traced by the tip of the vector as t increases from zero through 8π.

Vector functions can also be referred to in a different notation:

- $r(t) = \langle f(t), g(t) \rangle$ or
- $r(t) = \langle f(t), g(t), h(t) \rangle$

Properties

The domain of a vector-valued function is the intersection of the domain of the functions *f*, *g*, and *h*.

Derivative of a three-dimensional Vector Function

Many vector-valued functions, like scalar-valued functions, can be differentiated by simply differentiating the components in the Cartesian coordinate system.

Thus, if

$$r(t) = f(t)i + g(t)j + h(t)k$$

is a vector-valued function, then

$$\frac{dr(t)}{dt} = f'(t)i + g'(t)j + h'(t)k.$$

The vector derivative admits the following physical interpretation: if *r*(*t*) represents the position of a particle, then the derivative is the velocity of the particle

$$v(t) = \frac{dr(t)}{dt}.$$

Likewise, the derivative of the velocity is the acceleration

$$\frac{dv(t)}{dt} = a(t).$$

Partial Derivative

The partial derivative of a vector function a with respect to a scalar variable *q* is defined as

$$\frac{\partial a}{\partial q} = \sum_{i=1}^{3} \frac{\partial a_i}{\partial q} e_i$$

where a_i is the *scalar component* of a in the direction of e_i. It is also called the direction cosine of a and e_i or their dot product. The vectors e_1, e_2, e_3 form an orthonormal basis fixed in the reference frame in which the derivative is being taken.

Ordinary Derivative

If a is regarded as a vector function of a single scalar variable, such as time t, then the equation above reduces to the first ordinary time derivative of a with respect to t,

$$\frac{da}{dt} = \sum_{i=1}^{3} \frac{da_i}{dt} e_i .$$

Total Derivative

If the vector a is a function of a number n of scalar variables q_r $(r = 1,...,n)$, and each q_r is only a function of time t, then the ordinary derivative of a with respect to t can be expressed, in a form known as the total derivative, as

$$\frac{da}{dt} = \sum_{r=1}^{n} \frac{\partial a}{\partial q_r} \frac{dq_r}{dt} + \frac{\partial a}{\partial t} .$$

Some authors prefer to use capital D to indicate the total derivative operator, as in D/Dt. The total derivative differs from the partial time derivative in that the total derivative accounts for changes in a due to the time variance of the variables q_r.

Reference Frames

Whereas for scalar-valued functions there is only a single possible reference frame, to take the derivative of a vector-valued function requires the choice of a reference frame (at least when a fixed Cartesian coordinate system is not implied as such). Once a reference frame has been chosen, the derivative of a vector-valued function can be computed using techniques similar to those for computing derivatives of scalar-valued functions. A different choice of reference frame will, in general, produce a different derivative function. The derivative functions in different reference frames have a specific kinematical relationship.

Derivative of a Vector Function with Nonfixed Bases

The above formulas for the derivative of a vector function rely on the assumption that the basis vectors e_1,e_2,e_3 are constant, that is, fixed in the reference frame in which the derivative of a is being taken, and therefore the e_1,e_2,e_3 each has a derivative of identically zero. This often holds true for problems dealing with vector fields in a fixed coordinate system, or for simple problems in physics. However, many complex problems involve the derivative of a vector function in multiple moving reference frames, which means that the basis vectors will not necessarily be constant. In such a case where the basis vectors e_1, e_2, e_3 are fixed in reference frame **E**, but not in reference frame **N**, the more general formula for the ordinary time derivative of a vector in reference frame **N** is

$$\frac{{}^{N}da}{dt} = \sum_{i=1}^{3} \frac{da_i}{dt} e_i + \sum_{i=1}^{3} a_i \frac{{}^{N}de_i}{dt}$$

where the superscript N to the left of the derivative operator indicates the reference frame in which the derivative is taken. As shown previously, the first term on the right hand side is equal to the derivative of a in the reference frame where e_1, e_2, e_3 are constant, reference frame E. It also can be shown that the second term on the right hand side is equal to the relative angular velocity of the two reference frames cross multiplied with the vector a itself. Thus, after substitution, the formula relating the derivative of a vector function in two reference frames is

$$\frac{{}^{N}da}{dt} = \frac{{}^{E}da}{dt} + {}^{N}\omega^{E} \times a$$

where ${}^{N}\grave{u}^{E}$ is the angular velocity of the reference frame E relative to the reference frame N.

One common example where this formula is used is to find the velocity of a space-borne object, such as a rocket, in the inertial reference frame using measurements of the rocket's velocity relative to the ground. The velocity ${}^{N}v^{R}$ in inertial reference frame N of a rocket R located at position r^R can be found using the formula

$$\frac{{}^{N}d}{dt}(r^R) = \frac{{}^{E}d}{dt}(r^R) + {}^{N}\omega^{E} \times r^R.$$

where ${}^{N}\grave{u}^{E}$ is the angular velocity of the Earth relative to the inertial frame N. Since velocity is the derivative of position, ${}^{N}v^{R}$ and ${}^{E}v^{R}$ are the derivatives of r^R in reference frames N and E, respectively. By substitution,

$${}^{N}v^{R} = {}^{E}v^{R} + {}^{N}\omega^{E} \times r^{R}$$

where ${}^{E}v^{R}$ is the velocity vector of the rocket as measured from a reference frame E that is fixed to the Earth.

Derivative and Vector Multiplication

The derivative of the products of vector functions behaves similarly to the derivative of the products of scalar functions. Specifically, in the case of scalar multiplication of a vector, if p is a scalar variable function of q,

$$\frac{\partial}{\partial q}(pa) = \frac{\partial p}{\partial q}a + p\frac{\partial a}{\partial q}.$$

In the case of dot multiplication, for two vectors a and b that are both functions of q,

$$\frac{\partial}{\partial q}(a \cdot b) = \frac{\partial a}{\partial q} \cdot b + a \cdot \frac{\partial b}{\partial q}.$$

Similarly, the derivative of the cross product of two vector functions is

$$\frac{\partial}{\partial q}(a \times b) = \frac{\partial a}{\partial q} \times b + a \times \frac{\partial b}{\partial q}.$$

Derivative of an N-dimensional Vector Function

A function f of a real number t with values in the space R^n can be written as $f(t) = (f_1(t), f_2(t), \ldots, f_n(t))$. Its derivative equals

$$f'(t) = (f_{1'}(t), f_{2'}(t), \ldots, f_{n'}(t)).$$

If f is a function of several variables, say of $t \in R^m$, then the partial derivatives of the components of f form a $n \times m$ matrix called the *Jacobian matrix of f.*

Infinite-dimensional Vector Functions

If the values of a function f lie in an infinite-dimensional vector space X, such as a Hilbert space, then f may be called an *infinite-dimensional vector function.*

Functions with Values in a Hilbert Space

If the argument of f is a real number and X is a Hilbert space, then the derivative of f at a point t can be defined as in the finite-dimensional case:

$$f'(t) = \lim_{h \to 0} \frac{f(t+h) - f(t)}{h}.$$

Most results of the finite-dimensional case also hold in the infinite-dimensional case too, mutatis mutandis. Differentiation can also be defined to functions of several variables (e.g., $t \in R^n$ or even $t \in Y$, where Y is an infinite-dimensional vector space).

N.B. If X is a Hilbert space, then one can easily show that any derivative (and any other limit) can be computed componentwise: if

$$f = (f_1, f_2, f_3, \ldots)$$

(i.e., $f = f_1e_1 + f_2e_2 + f_3e_3 + \cdots$, where $e_1, e_2, e_3, \ldots$ is an orthonormal basis of the space X), and $f'(t)$ exists, then

$$f'(t) = (f_{1'}(t), f_{2'}(t), f_{3'}(t), \ldots).$$

However, the existence of a componentwise derivative does not guarantee the existence of a derivative, as componentwise convergence in a Hilbert space does not guarantee convergence with respect to the actual topology of the Hilbert space.

Other infinite-dimensional vector spaces

Most of the above hold for other topological vector spaces X too. However, not as many classical results hold in the Banach space setting, e.g., an absolutely continuous function with values in a suitable Banach space need not have a derivative anywhere. Moreover, in most Banach spaces setting there are no orthonormal bases.

Integral Transform

In mathematics, an integral transform is any transform T of the following form:

$$(Tf)(u) = \int_{t_1}^{t_2} K(t,u)\, f(t)\, dt$$

The input of this transform is a function f, and the output is another function Tf. An integral transform is a particular kind of mathematical operator. There are numerous useful integral transforms. Each is specified by a choice of the function K of two variables, the kernel function or nucleus of the transform.

Some kernels have an associated *inverse kernel* $K^{-1}(u, t)$ which (roughly speaking) yields an inverse transform:

$$f(t) = \int_{u_1}^{u_2} K^{-1}(u,t)(Tf(u))du$$

A *symmetric kernel* is one that is unchanged when the two variables are permuted.

Motivation

Mathematical notation aside, the motivation behind integral transforms is easy to understand. There are many classes of problems that are difficult to solve—or at least quite unwieldy algebraically—in their original representations. An integral transform "maps" an equation from its original "domain" into another domain. Manipulating and solving the equation in the target domain can be much easier than manipulation and solution in the original domain. The solution is then mapped back to the original domain with the inverse of the integral transform.

History

The precursor of the transforms were the Fourier series to express functions in finite intervals. Later the Fourier transform was developed to remove the requirement of finite intervals. Using the Fourier series, just about any practical function of time (the voltage across the terminals of an electronic device for example) can be represented as a sum of sines and cosines, each suitably scaled (multiplied by a constant factor), shifted (advanced or retarded in time) and "squeezed" or "stretched" (increasing or decreasing the frequency). The sines and cosines in the Fourier series are an example of an orthonormal basis.

Importance of Orthogonality

The individual basis functions have to be orthogonal. That is, the product of two dissimilar basis functions—integrated over their domain—must be zero. An integral transform, in actuality, just changes the representation of a function from one orthogonal basis to another. Each point in the representation of the transformed function in the image of the transform corresponds to the contribution of a given orthogonal basis function to the expansion. The process of expanding a function from its "standard" representation to a sum of a number of orthonormal basis functions, suitably scaled and shifted, is termed "spectral factorization." This is similar in concept to the description of a point in space in terms of three discrete components, namely, its x, y, and z coordinates. Each axis correlates only to itself and nothing to the other orthogonal axes. Note the terminological consistency: the determination of the amount by which an individual orthonormal basis function must be scaled in the spectral factorization of a function, F, is termed the "projection" of F onto that basis function.

The normal Cartesian graph *per se* of a function can be thought of as an orthonormal expansion. Indeed, each point just reflects the contribution of a given orthonormal basis function to the sum. In this way, the graph of a continuous real-valued function in the plane corresponds to an *infinite* set of basis functions; if the number of basis functions were finite, the curve would consist of a discrete set of points rather than a continuous contour.

Usage Example

As an example of an application of integral transforms, consider the Laplace transform. This is a technique that maps differential or integro-differential equations in the "time" domain into polynomial equations in what is termed the "complex frequency" domain. (Complex frequency is similar to actual, physical frequency but rather more general.

Specifically, the imaginary component ω of the complex frequency $s = -\sigma + i\omega$ corresponds to the usual concept of frequency, *viz.*, the speed at which a sinusoid cycles, whereas the real component σ of the complex frequency corresponds to the degree of "damping".)

The equation cast in terms of complex frequency is readily solved in the complex frequency domain (roots of the polynomial equations in the complex frequency domain correspond to eigenvalues in the time domain), leading to a "solution" formulated in the frequency domain. Employing the inverse transform, *i.e.*, the inverse procedure of the original Laplace transform, one obtains a time-domain solution. In this example, polynomials in the complex frequency domain (typically occurring in the denominator) correspond to power series in the time domain, while axial shifts in the complex frequency domain correspond to damping by decaying exponentials in the time domain.

The Laplace transform finds wide application in physics and particularly in electrical engineering, where the characteristic equations that describe the behaviour of an electric circuit in the complex frequency domain correspond to linear combinations of exponentially damped, scaled, and time-shifted sinusoids in the time domain. Other integral transforms find special applicability within other scientific and mathematical disciplines.

Different Domains

Here integral transforms are defined for functions on the real numbers, but they can be defined more generally for functions on a group.

- If instead one uses functions on the circle (periodic functions), integration kernels are then biperiodic functions; convolution by functions on the circle yields circular convolution.
- If one uses functions on the cyclic group of order n (C_n or Z/nZ), one obtains $n \times n$ matrices as integration kernels; convolution corresponds to circulant matrices.

General Theory

Although the properties of integral transforms vary widely, they have some properties in common. For example, every integral transform is a linear operator, since the integral is a linear operator, and in fact if the kernel is allowed to be a generalized function then all linear operators are integral transforms (a properly formulated version of this statement is the Schwartz kernel theorem).

The general theory of such integral equations is known as Fredholm theory. In this theory, the kernel is understood to be a compact operator acting on a Banach space of functions. Depending on the situation, the kernel is then variously referred to as the Fredholm operator, the nuclear operator or the Fredholm kernel.

Chapter 8

Structural Dynamics

Structural analysis is mainly concerned with finding out the behaviour of a structure when subjected to some action. This action can be in the form of load due to the weight of things such as people, furniture, wind, snow, etc. or some other kind of excitation such as an earthquake, shaking of the ground due to a blast nearby, etc. In essence all these loads are dynamic, including the self-weight of the structure because at some point in time these loads were not there. The distinction is made between the dynamic and the static analysis on the basis of whether the applied action has enough acceleration in comparison to the structure's natural frequency.

If a load is applied sufficiently slowly, the inertia forces (Newton's second law of motion) can be ignored and the analysis can be simplified as static analysis. Structural dynamics, therefore, is a type of structural analysis which covers the behaviour of structures subjected to dynamic (actions having high acceleration) loading. Dynamic loads include people, wind, waves, traffic, earthquakes, and blasts. Any structure can be subject to dynamic loading. Dynamic analysis can be used to find dynamic displacements, time history, and modal analysis.

A dynamic analysis is also related to the inertia forces developed by a structure when it is excited by means of dynamic loads applied suddenly (e.g., wind blasts, explosion, earthquake). A static load is one which varies very slowly. A dynamic load is one which changes with time fairly quickly in comparison to the structure's natural frequency. If it changes slowly, the structure's response may be determined with static analysis, but if it varies quickly (relative to the structure's ability to respond), the response must be determined with a dynamic analysis. Dynamic analysis for simple structures can be carried out manually, but for complex structures finite element analysis can be used to calculate the mode shapes and frequencies. An open-source, lightweight, free software DYSSOLVE can be used to solve basic structural dynamics problems.

Displacements: A dynamic load can have a significantly larger effect than a static load of the same magnitude due to the structure's inability to respond quickly to the loading (by deflecting). The increase in the effect of a dynamic load is given by the dynamic amplification factor (DAF):

$$DAF = \frac{u_{max}}{u_{static}}$$

where u is the deflection of the structure due to the applied load.

Graphs of dynamic amplification factors vs non-dimensional rise time (t_r/T) exist for standard loading functions. Hence the DAF for a given loading can be read from the graph, the static deflection can be easily calculated for simple structures and the dynamic deflection found.

Time History Analysis

A full time history will give the response of a structure over time during and after the application of a load. To find the full time history of a structure's response, you must solve the structure's equation of motion.

Example

A simple single degree of freedom system (a mass, M, on a spring of stiffness, k for example) has the following equation of motion:

$$M\ddot{x} + kx = F(t)$$

where $\ddot{x}$ is the acceleration (the double derivative of the displacement) and x is the displacement. If the loading $F(t)$ is a Heaviside step function (the sudden application of a constant load), the solution to the equation of motion is:

$$x = \frac{F_0}{k}[1 - cos(\omega t)]$$

where $\omega = \sqrt{\frac{k}{M}}$ and the fundamental natural frequency, $f = \frac{\omega}{2\pi}$.

The static deflection of a single degree of freedom system is:

$$x_{static} = \frac{F_0}{k}$$

so you can write, by combining the above formulae:

$$x = x_{static}[1 - cos(\omega t)]$$

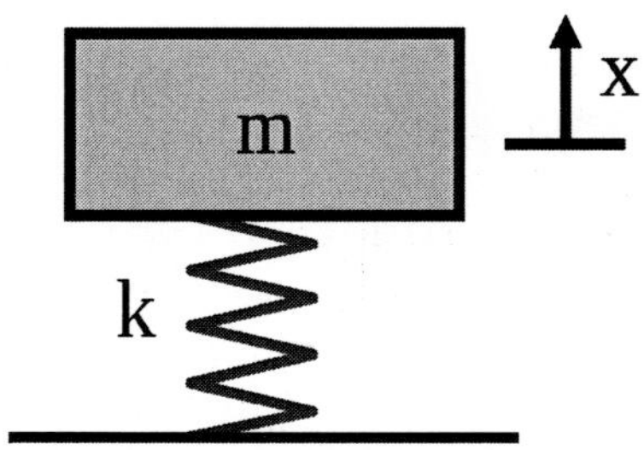

This gives the (theoretical) time history of the structure due to a load $F(t)$, where the false assumption is made that there is no damping.

Although this is too simplistic to apply to a real structure, the Heaviside Step Function is a reasonable model for the application of many real loads, such as the sudden addition of a piece of furniture, or the removal of a prop to a newly cast concrete floor. However, in reality loads are never applied instantaneously - they build up over a period of time (this may be very short indeed). This time is called the rise time.

As the number of degrees of freedom of a structure increases it very quickly becomes too difficult to calculate the time history manually - real structures are analysed using non-linear finite element analysis software.

Damping

Any real structure will dissipate energy (mainly through friction). This can be modelled by modifying the DAF:

$$DAF = 1 + e^{-c\pi}$$

where $c = \dfrac{\text{Damping Coefficient}}{\text{Critical Damping Coefficient}}$ and is typically 2%-10% depending on the type of construction:

- Bolted steel ~6%
- Reinforced concrete ~ 5%
- Welded steel ~ 2%

Generally damping would be ignored for non-transient events (such as wind loading or crowd loading), but would be important for transient events (for example, an impulse load such as a bomb blast).

Modal Analysis

A modal analysis calculates the frequency modes or natural frequencies of a given system, but not necessarily its full time history response to a given input. The natural frequency of a system is dependent only on the stiffness of the structure and the mass which participates with the structure (including self-weight). It is not dependent on the load function.

It is useful to know the modal frequencies of a structure as it allows you to ensure that the frequency of any applied periodic loading will not coincide with a modal frequency and hence cause resonance, which leads to large oscillations.

The method is:

1. Find the natural modes (the shape adopted by a structure) and natural frequencies
2. Calculate the response of each mode
3. Optionally superpose the response of each mode to find the full modal response to a given loading

Energy Method

It is possible to calculate the frequency of different mode shape of system manually by the energy method. For a given mode shape of a multiple degree of freedom system you can find an "equivalent– mass, stiffness and applied force for a single degree of freedom system.

For simple structures the basic mode shapes can be found by inspection, but it is not a conservative method. Rayleigh's principle states:

"The frequency ω of an arbitrary mode of vibration, calculated by the energy method, is always greater than - or equal to - the fundamental frequency ω_n."

For an assumed mode shape $\overline{u}(x)$, of a structural system with mass M; stiffness, EI (Young's modulus, E, multiplied by the second moment of area, I); and applied force, *F*(*x*):

$$\text{Equivalent mass, } M_{eq} = \int M\overline{u}^2 du$$

$$\text{Equivalent stiffness, } k_{eq} = \int EI(\frac{d^2\overline{u}}{dx^2})dx$$

$$\text{Equivalent force, } F_{eq} = \int F\overline{u}dx$$

then, as above:

$$\omega = \sqrt{\frac{k_{eq}}{M_{eq}}}$$

Modal Response

The complete modal response to a given load F(x,t) is $v(x,t) = \sum u_n(x,t)$. The summation can be carried out by one of three common methods:

- Superpose complete time histories of each mode (time consuming, but exact)
- Superpose the maximum amplitudes of each mode (quick but conservative)
- Superpose the square root of the sum of squares (good estimate for well-separated frequencies, but unsafe for closely spaced frequencies)

To superpose the individual modal responses manually, having calculated them by the energy method:

Assuming that the rise time t_r is known ($T = 2\pi/\omega$), it is possible to read the DAF from a standard graph. The static displacement can be calculated with $u_{static} = \frac{F_{1,eq}}{k_{1,eq}}$. The dynamic displacement for the chosen mode and applied force can then be found from:

$$u_{max} = u_{static} DAF$$

Modal Participation Factor

For real systems there is often mass participating in the forcing function (such as the mass of ground in an earthquake) and mass participating in inertia effects (the mass of the structure itself, M_{eq}). The modal participation factor Γ is a comparison of these two masses. For a single degree of freedom system Γ = 1.

$$\Gamma = \frac{\sum M_n \bar{u}_n}{\sum M_n \bar{u}_n^2}$$

Transient Friction Loading

Transient friction loading, also known as TFL, is the mechanical stress induced on an object due to transient or vibrational frictional forces.

Examples

A classic example of TFL is the wooden block sliding over an unlevel, non-planar surface. Due to the transient response of the contact force, the resultant frictional force is noisy. The induced stress is concisely described as TFL.

Another example of TFL is the internal stresses on a hydraulic ram operating in a vibrational environment. Due to the oscilliatory nature of the acceleration experienced by the ram, in n-dimensions, the resulting frictional response is described as transient friction loading.

Damping

In physics, damping is an effect that reduces the amplitude of oscillations in an oscillatory system (except for mass-dominated systems where $\omega / \omega_0 > - 2$), particularly the harmonic oscillator.

This effect is linearly related to the velocity of the oscillations. This restriction leads to a linear differential equation of motion, and a simple analytic solution.

In mechanics, damping may be realised using a dashpot. This device uses the viscous drag of a fluid, such as oil, to provide a resistance that is related linearly to velocity. The damping force F_c is expressed as follows:

$$F_c = -c\frac{dx}{dt},$$

where c is the *viscous damping coefficient*, given in units of newton seconds per meter (N s/m) or simply kilograms per second. In engineering applications it is often desirable to linearise non-linear drag forces. This may by finding an equivalent work coefficient in the case of harmonic forcing. In non-harmonic cases, restrictions on the speed may lead to accurate linearisation.

Generally, damped harmonic oscillators satisfy the second-order differential equation:

$$\frac{d^2x}{dt^2} + 2\zeta\omega_0\frac{dx}{dt} + \omega_0^2 x = 0,$$

where ω_0 is the undamped angular frequency of the oscillator and ζ is a constant called the damping ratio.

The value of the damping ratio ζ determines the behaviour of the system. A damped harmonic oscillator can be:

- *Overdamped* ($\zeta > 1$): The system returns (exponentially decays) to equilibrium without oscillating. Larger values of the damping ratio ζ return to equilibrium more slowly.
- *Critically damped* ($\zeta = 1$): The system returns to equilibrium as quickly as possible without oscillating. This is often desired for the damping of systems such as doors.
- *Underdamped* ($0 < \zeta < 1$): The system oscillates (at reduced frequency compared to the *undamped* case) with the amplitude gradually decreasing to zero.
- *Undamped* ($\zeta = 0$): The system oscillates at its natural resonant frequency (ω_0).

Definition

In physics and engineering, damping may be mathematically modelled as a force synchronous with the velocity of the object but opposite in direction to it. If such force is also proportional to the velocity, as for a simple mechanical viscous damper (dashpot), the force F may be related to the velocity v by

$$F = -cv,$$

where c is the *damping coefficient*, given in units of newton-seconds per meter.

This force may be used as an approximation to the friction caused by drag. While friction is related to v^2, if the velocity is restricted a small range, this non-linear effect may be small. In such a situation, a linearised friction coefficient c_{lin} may be determined which produces little error compared with the second order solution.

This relationship is perfectly analogous to electrical resistance as described by Ohm's law.

Example: mass–spring–damper

An ideal mass–spring–damper system with mass m, spring constant k and viscous damper of damping coefficient c is subject to an oscillatory force

$$F_s = -kx$$

and a damping force

$$F_d = -cv = -c\frac{dx}{dt} = -c\dot{x}.$$

The values can be in any consistent system of units; for example, in SI units, m in kilograms, k in newtons per meter, and c in newton-seconds per meter or kilograms per second.

Treating the mass as a free body and applying Newton's second law, the total force F_{tot} on the body is

$$F_{tot} = ma = m\frac{d^2x}{dt^2} = m\ddot{x}.$$

where a is the acceleration of the mass and x is the displacement of the mass relative to a fixed point of reference.

Since $F_{tot} = F_s + F_d$,

$$m\ddot{x} = -kx + -c\dot{x}.$$

This differential equation may be rearranged into

$$\ddot{x} + \frac{c}{m}\dot{x} + \frac{k}{m}x = 0.$$

The following parameters are then defined:

$$\omega_0 = \sqrt{\frac{k}{m}}$$

$$\zeta = \frac{c}{2\sqrt{mk}}.$$

The first parameter, ω_0, is called the (undamped) natural frequency of the system. The second parameter, ζ, is called the *damping ratio*. The natural frequency represents an angular frequency, expressed in radians per second. The damping ratio is a dimensionless quantity.

The differential equation now becomes

$$\ddot{x} + 2\zeta\omega_0\dot{x} + \omega_0^2 x = 0.$$

Continuing, we can solve the equation by assuming a solution x such that:

$$x = e^{\gamma t}$$

where the parameter γ (gamma) is, in general, a complex number.

Substituting this assumed solution back into the differential equation gives

$$\gamma^2 + 2\zeta\omega_0\gamma + \omega_0^2 = 0,$$

which is the characteristic equation.

Solving the characteristic equation will give two roots, γ_+ and γ_- The solution to the differential equation is thus

$$x(t) = Ae^{\gamma_+ t} + Be^{\gamma_- t},$$

where A and B are determined by the initial conditions of the system:

$$A = x(0) + \frac{\gamma_+ x(0) - \dot{x}(0)}{\gamma_- - \gamma_+}$$

$$B = -\frac{\gamma_+ x(0) - \dot{x}(0)}{\gamma_- - \gamma_+}.$$

System Behaviour

The behaviour of the system depends on the relative values of the two fundamental parameters, the natural frequency ω_0

and the damping ratio ζ. In particular, the qualitative behaviour of the system depends crucially on whether the quadratic equation for γ has one real solution, two real solutions, or two complex conjugate solutions.

Critical Damping (ζ = 1)

When $\zeta = 1$, there is a double root γ (defined above), which is real. The system is said to be *critically damped*.

A critically damped system converges to zero as fast as possible without oscillating. An example of critical damping is the door closer seen on many hinged doors in public buildings.

The recoil mechanisms in most guns are also critically damped so that they return to their original position, after the recoil due to firing, in the least possible time.

In this case, with only one root γ, there is in addition to the solution $x(t) = e^{\gamma t}$ a solution $x(t) = te^{\gamma t}$:

$$x(t) = (A + Bt)e^{-\omega_0 t}$$

where A and B are determined by the initial conditions of the system (usually the initial position and velocity of the mass):

$$A = x(0)$$

$$B = \dot{x}(0) + \omega_0 x(0)$$

Over-damping (ζ > 1)

When $\zeta > 1$, the system is *over-damped* and there are two different real roots.

An over-damped door-closer will take longer to close than a critically damped door would.

The solution to the motion equation is:

$$x(t) = Ae^{\gamma_+ t} + Be^{\gamma_- t}$$

where A and B are determined by the initial conditions of the system:

$$A = x(0) + \frac{\gamma_+ x(0) - \dot{x}(0)}{\gamma_- - \gamma_+}$$

$$B = -\frac{\gamma_+ x(0) - \dot{x}(0)}{\gamma_- - \gamma_+}.$$

Under-damping (0 d– ζ < 1)

Finally, when $0 < \zeta < 1$, γ is complex, and the system is *under-damped*. In this situation, the system will oscillate at the natural damped frequency ω_d, which is a function of the natural frequency and the damping ratio. To continue the analogy, an underdamped door closer would close quickly, but would hit the door frame with significant velocity, or would oscillate in the case of a swinging door.

In this case, the solution can be generally written as:

$$x(t) = e^{-\zeta\omega_0 t}(A\cos(\omega_d t) + B\sin(\omega_d t))$$

where

$$\omega_d = \omega_0\sqrt{1-\zeta^2}$$

represents the *damped frequency* or *ringing frequency* of the system, and A and B are again determined by the initial conditions of the system:

$$A = x(0)$$

$$B = \frac{1}{\omega_d}(\zeta\omega_0 x(0) + \dot{x}(0)).$$

This "damped frequency" is not to be confused with the *damped resonant frequency* or *peak frequency* ω_{peak}. This is the frequency at which a moderately underdamped (ζ < 1/–2) simple 2nd-order harmonic oscillator has its maximum gain (or peak transmissibility) when driven by a sinusoidal input. The frequency at which this peak occurs is given by:

$$\omega_{peak} = \omega_0\sqrt{1-2\zeta^2}\ .$$

For an under-damped system, the value of ζ can be found by examining the logarithm of the ratio of succeeding amplitudes of a system. This is called the logarithmic decrement.

Alternative Models

Viscous damping models, although widely used, are not the only damping models. A wide range of models can be found in specialized literature. One is the so called "hysteretic damping model" or "structural damping model". When a metal beam is vibrating, the internal damping can be better described by a force proportional to the displacement but in phase with the velocity. In such case, the differential equation that describes the free movement of a single-degree-of-freedom system becomes:

$$m\ddot{x} + hxi + kx = 0$$

where h is the hysteretic damping coefficient and i denotes the imaginary unit; the presence of i is required to synchronise the damping force to the velocity (xi being in phase with the velocity). This equation is more often written as:

$$m\ddot{x} + k(1 + i\eta)x = 0$$

where η is the hysteretic damping ratio, that is, the fraction of energy lost in each cycle of the vibration.

Although requiring complex analysis to solve the equation, this model reproduces the real behaviour of many vibrating structures more closely than the viscous model.

A more general model that also requires complex analysis, the fractional model not only includes both the viscous and hysteretic models, but also allows for intermediate cases (useful for some polymers):

$$m\ddot{x} + A\frac{d^r x}{dt^r}i + kx = 0$$

where r is any number, usually between 0 (for hysteretic) and 1 (for viscous), and A is a general damping (h for hysteretic and c for viscous) coefficient.

Resonance

In physics, resonance is the tendency of a system to oscillate with greater amplitude at some frequencies than at others. Frequencies at which the response amplitude is a relative maximum are known as the system's resonant frequencies, or resonance frequencies. At these frequencies, even small periodic driving forces can produce large amplitude oscillations, because the system stores vibrational energy.

Resonance occurs when a system is able to store and easily transfer energy between two or more different storage modes (such as kinetic energy and potential energy in the case of a pendulum). However, there are some losses from cycle to cycle, called damping. When damping is small, the resonant frequency is approximately equal to the natural frequency of the system, which is a frequency of unforced vibrations. Some systems have multiple, distinct, resonant frequencies.

Resonance phenomena occur with all types of vibrations or waves: there is mechanical resonance, acoustic resonance, electromagnetic resonance, nuclear magnetic resonance (NMR), electron spin resonance (ESR) and resonance of quantum wave functions. Resonant systems can be used to generate vibrations of a specific frequency (e.g. musical instruments), or pick out specific frequencies from a complex vibration containing many frequencies (e.g. filters).

Resonance was recognised by Galileo Galilei with his investigations of pendulums and musical strings beginning in 1602.

Examples

One familiar example is a playground swing, which acts as a pendulum. Pushing a person in a swing in time with the natural interval of the swing (its resonant frequency) will make the swing go higher and higher (maximum amplitude), while attempts to push the swing at a faster or slower tempo will result in smaller arcs. This is because the energy the swing absorbs is maximised when the pushes are 'in phase' with the swing's oscillations, while some of the swing's energy is actually extracted by the opposing force of the pushes when they are not.

Resonance occurs widely in nature, and is exploited in many man-made devices. It is the mechanism by which virtually all sinusoidal waves and vibrations are generated. Many sounds we hear, such as when hard objects of metal, glass, or wood are struck, are caused by brief resonant vibrations in the object. Light and other short wavelength electromagnetic radiation is produced by resonance on an atomic scale, such as electrons in atoms. Other examples are.

Mechanical and acoustic resonance:

- the timekeeping mechanisms of modern clocks and watches, e.g. the balance wheel in a mechanical watch and the quartz crystal in a quartz watch

- the tidal resonance of the Bay of Fundy
- acoustic resonances of musical instruments and human vocal cords
- the shattering of a crystal wineglass when exposed to a musical tone of the right pitch (its resonant frequency)

Electrical resonance:

- electrical resonance of tuned circuits in radios and TVs that allow radio frequencies to be selectively received

Optical resonance:

- creation of coherent light by optical resonance in a laser cavity

Orbital resonance in astronomy:

- orbital resonance as exemplified by some moons of the solar system's gas giants

Atomic, Particle, and Molecular Resonance

- material resonances in atomic scale are the basis of several spectroscopic techniques that are used in condensed matter physics.
- Nuclear magnetic resonance
- Mössbauer effect
- Electron spin resonance.

Theory

The exact response of a resonance, especially for frequencies far from the resonant frequency, depends on the details of the physical system, and is usually not exactly symmetric about the resonant frequency, as illustrated for the simple harmonic oscillator above. For a lightly damped linear oscillator with a resonance frequency Ω, the *intensity* of oscillations I when the system is driven with a driving frequency ω is typically approximated by a formula that is symmetric about the resonance frequency:

$$I(\omega) \propto \frac{\left(\frac{\Gamma}{2}\right)^2}{(\omega - \Omega)^2 + \left(\frac{\Gamma}{2}\right)^2}.$$

The intensity is defined as the square of the amplitude of the oscillations. This is a Lorentzian function, and this response is found in many physical situations involving resonant systems. Γ is a parameter dependent on the damping of the oscillator, and is known as the *linewidth* of the resonance. Heavily damped oscillators tend to have broad linewidths, and respond to a wider range of driving frequencies around the resonant frequency. The linewidth is inversely proportional to the Q factor, which is a measure of the sharpness of the resonance.

In electrical engineering, this approximate symmetric response is known as the *universal resonance curve*, a concept introduced by Frederick E. Terman in 1932 to simplify the approximate analysis of radio circuits with a range of centre frequencies and Q values.

Resonators

A physical system can have as many resonant frequencies as it has degrees of freedom; each degree of freedom can vibrate as a harmonic oscillator. Systems with one degree of freedom, such as a mass on a spring, pendulums, balance wheels, and LC tuned circuits have one resonant frequency. Systems with two degrees of freedom, such as coupled pendulums and resonant transformers can have two resonant frequencies. As the number of coupled harmonic oscillators grows, the time it takes to transfer energy from one to the next becomes significant. The vibrations in them begin to travel through the coupled harmonic oscillators in waves, from one oscillator to the next.

Extended objects that experience resonance due to vibrations inside them are called resonators, such as organ pipes, vibrating strings, quartz crystals, microwave cavities, and laser rods. Since these can be viewed as being made of millions of coupled moving parts (such as atoms), they can have millions of resonant frequencies.

The vibrations inside them travel as waves, at an approximately constant velocity, bouncing back and forth between the sides of the resonator. If the distance between the sides is d, the length of a round trip is $2d$. In order to cause resonance, the phase of a sinusoidal wave after a round trip has to be equal to the initial phase, so the waves will reinforce.

So the condition for resonance in a resonator is that the round trip distance, , be equal to an integer number of wavelengths of the wave:

$$2d = N\lambda, \qquad N \in \{1,2,3,\ldots\}$$

If the velocity of a wave is v, the frequency is $f = v / \lambda$ so the resonant frequencies are:

$$f = \frac{Nv}{2d} \qquad N \in \{1,2,3,\ldots\}$$

So the resonant frequencies of resonators, called normal modes, are equally spaced multiples of a lowest frequency called the fundamental frequency. The multiples are often called overtones. There may be several such series of resonant frequencies, corresponding to different modes of vibration.

Q Factor

In physics and engineering the quality factor or Q factor is a dimensionless parameter that describes how under-damped an oscillator or resonator is, or equivalently, characterises a resonator's bandwidth relative to its centre frequency. Higher Q indicates a lower rate of energy loss relative to the stored energy of the resonator; the oscillations die out more slowly. A pendulum suspended from a high-quality bearing, oscillating in air, has a high Q, while a pendulum immersed in oil has a low one. Resonators with high quality factors have low damping so that they ring longer.

Explanation

Sinusoidally driven resonators having higher Q factors resonate with greater amplitudes (at the resonant frequency) but have a smaller range of frequencies around that frequency for which they resonate; the range of frequencies for which the oscillator resonates is called the bandwidth. Thus, a high-Q tuned circuit in a radio receiver would be more difficult to tune, but would have more selectivity; it would do a better job of filtering out signals from other stations that lie nearby on the spectrum. High-Q oscillators oscillate with a smaller range of frequencies and are more stable.

The quality factor of oscillators varies substantially from system to system. Systems for which damping is important (such as dampers keeping a door from slamming shut) have Q near ½. Clocks, lasers, and other resonating systems that need either strong resonance or high frequency stability have high quality factors. Tuning forks have quality factors around 1000. The quality factor of atomic clocks, superconducting RF cavities used in accelerators, and some high-Q lasers can reach as high as 10 and higher.

There are many alternative quantities used by physicists and engineers to describe how damped an oscillator is. Important examples include: the damping ratio, relative bandwidth, linewidth and bandwidth measured in octaves.

The concept of "Q" originated with K.S. Johnson of Western Electric Company's Engineering Department while evaluating the quality of coils (inductors). His choice of the symbol Q was only because all other letters of the alphabet were taken. The term was not intended as an abbreviation for "quality" or "quality factor", although these terms have grown to be associated with it.

Definition of the Quality Factor

In the context of resonators, Q is defined in terms of the ratio of the energy stored in the resonator to the energy supplied by a generator, per cycle, to keep signal amplitude constant, at a frequency (the resonant frequency), f_r, where the stored energy is constant with time:

$$Q = 2\pi \times \frac{\text{Energy Stored}}{\text{Energy dissipated per cycle}} = 2\pi f_r \times \frac{\text{Energy Stored}}{\text{Power Loss}}.$$

The factor 2π makes Q expressible in simpler terms, involving only the coefficients of the second-order differential equation describing most resonant systems, electrical or mechanical. In electrical systems, the stored energy is the sum of energies stored in lossless inductors and capacitors; the lost energy is the sum of the energies dissipated in resistors per cycle. In mechanical systems, the stored energy is the maximum possible stored energy, or the total energy, i.e. the sum of the potential and kinetic energies at some point in time; the lost energy is the work done by an external conservative force, per cycle, to maintain amplitude.

For high values of Q, the following definition is also mathematically accurate:

$$Q = \frac{f_r}{\Delta f} = \frac{\omega_r}{\Delta\omega},$$

where f_r is the resonant frequency, Δf is the half-power bandwidth i.e. the bandwidth over which the power of vibration is greater than half the power at the resonant frequency, $\omega_r = 2\pi f_r$ is the angular resonant frequency, and $\Delta\omega$ is the angular half-power bandwidth.

More generally and in the context of reactive component specification (especially inductors), the frequency-dependent definition of Q is used:

$$Q(\omega) = \omega \times \frac{\text{Maximum Energy Stored}}{\text{Power Loss}},$$

where ω is the angular frequency at which the stored energy and power loss are measured. This definition is consistent with its usage in describing circuits with a single reactive element (capacitor or inductor), where it can be shown to be equal to the ratio of reactive power to real power.

Q Factor and Damping

The Q factor determines the qualitative behaviour of simple damped oscillators.

- A system with low quality factor ($Q < ½$) is said to be overdamped. Such a system doesn't oscillate at all, but when displaced from its equilibrium steady-state output it returns to it by exponential decay, approaching the steady state value asymptotically. It has an impulse response that is the sum of two decaying exponential functions with different rates of decay.

  As the quality factor decreases the slower decay mode becomes stronger relative to the faster mode and dominates the system's response resulting in a slower system. A second-order low-pass filter with a very low quality factor has a nearly first-order step response; the system's output responds to a step input by slowly rising toward an asymptote.

- A system with high quality factor ($Q > ½$) is said to be underdamped. Underdamped systems combine oscillation at a specific frequency with a decay of the amplitude of the signal. Underdamped systems with a low quality factor (a little above $Q = ½$) may oscillate only once or a few times before dying out. As the quality factor increases, the relative amount of damping decreases. A high-quality bell rings with a single pure tone for a very long time after being struck. A purely oscillatory system, such as a bell that rings forever, has an infinite quality factor. More generally, the output of a second-order low-pass filter with a very high quality factor responds to a step input by quickly rising above, oscillating around, and eventually converging to a steady-state value.
- A system with an intermediate quality factor ($Q = ½$) is said to be critically damped. Like an overdamped system, the output does not oscillate, and does not overshoot its steady-state output (i.e., it approaches a steady-state asymptote). Like an underdamped response, the output of such a system responds quickly to a unit step input. Critical damping results in the fastest response (approach to the final value) possible without overshoot. Real system specifications usually allow some overshoot for a faster initial response or require a slower initial response to provide a safety margin against overshoot.

In negative feedback systems, the dominant closed-loop response is often well-modelled by a second-order system. The phase margin of the open-loop system sets the quality factor Q of the closed-loop system; as the phase margin decreases, the approximate second-order closed-loop system is made more oscillatory (i.e., has a higher quality factor).

Quality Factors of Common Systems:

- A unity gain Sallen–Key filter topology with equivalent capacitors and equivalent resistors is critically damped (i.e., $Q = 1/2$).
- A second order Butterworth filter (i.e., continuous-time filter with the flattest passband frequency response) has an underdamped $Q = 1/\sqrt{2}$.

- A Bessel filter (i.e., continuous-time filter with flattest group delay) has an underdamped $Q = 1/\sqrt{3}$.

Physical Interpretation of Q

Physically speaking, Q is 2π times the ratio of the total energy stored divided by the energy lost in a single cycle or equivalently the ratio of the stored energy to the energy dissipated over one radian of the oscillation.

It is a dimensionless parameter that compares the time constant for decay of an oscillating physical system's amplitude to its oscillation period. Equivalently, it compares the frequency at which a system oscillates to the rate at which it dissipates its energy.

Equivalently (for large values of Q), the Q factor is approximately the number of oscillations required for a freely oscillating system's energy to fall off to $1/e^{2\pi}$, or about 1/535 or 0.2%, of its original energy.

The width (bandwidth) of the resonance is given by

$$\Delta f = \frac{f_0}{Q},$$

where f_0 is the resonant frequency, and Δf, the bandwidth, is the width of the range of frequencies for which the energy is at least half its peak value.

The factors Q, damping ratio ζ, and attenuation α are related such that

$$\zeta = \frac{1}{2Q} = \frac{\alpha}{\omega_0}.$$

So the quality factor can be expressed as

$$Q = \frac{1}{2\zeta} = \frac{\omega_0}{2\alpha},$$

and the exponential attenuation rate can be expressed as

$$\alpha = \zeta\omega_0 = \frac{\omega_0}{2Q}.$$

For any 2nd order low-pass filter, the response function of the filter is

$$H(s) = \frac{\omega_0^2}{s^2 + \underbrace{\frac{\omega_0}{Q}}_{2\zeta\omega_0 = 2\alpha} s + \omega_0^2}$$

For this system, $Q > 0.5$ when (i.e., when the system is underdamped), it has two complex conjugate poles that each have a real part of α. That is, the attenuation parameter represents the rate of exponential decay of the oscillations (e.g., after an impulse) of the system. A higher quality factor implies a lower attenuation, and so high Q systems oscillate for long times. For example, high quality bells have an approximately pure sinusoidal tone for a long time after being struck by a hammer.

Electrical Systems

For an electrically resonant system, the Q factor represents the effect of electrical resistance and, for electromechanical resonators such as quartz crystals, mechanical friction.

RLC Circuits

In an ideal series RLC circuit, and in a tuned radio frequency receiver (TRF) the Q factor is:

$$Q = \frac{1}{R}\sqrt{\frac{L}{C}},$$

where R, L and C are the resistance, inductance and capacitance of the tuned circuit, respectively. The larger the series resistance, the lower the circuit Q.

For a parallel RLC circuit, the Q factor is the inverse of the series case:

$$Q = R\sqrt{\frac{C}{L}},$$

Consider a circuit where R, L and C are all in parallel. The lower the parallel resistance, the more effect it will have in damping the circuit and thus the lower the Q. This is useful in filter design to determine the bandwidth.

In a parallel LC circuit where the main loss is the resistance of the inductor, R, in series with the inductance, L, Q is as in the series circuit. This is a common circumstance for resonators, where limiting the resistance of the inductor to improve Q and narrow the bandwidth is the desired result.

Individual Reactive Components

The Q of an individual reactive component depends on the frequency at which it is evaluated, which is typically the resonant frequency of the circuit that it is used in.

The Q of an inductor with a series loss resistance is the same as the Q of a resonant circuit using that inductor with a perfect capacitor:

$$Q_L = \frac{X_L}{R_L} = \frac{\omega L}{R_L}$$

Where:

- ω is frequency in radians per second,
- L is the inductance,
- X_L is the inductive reactance, and
- R_L is the series resistance of the inductor.

The Q of a capacitor with a series loss resistance is the same as the Q of a resonant circuit using that capacitor with a perfect inductor:

$$Q_C = \frac{X_C}{R_C} = \frac{1}{\omega C R_C}$$

Where:

- ω is frequency in radians per second,
- C is the capacitance,
- X_C is the capacitive reactance, and
- R_C is the series resistance of the capacitor.

In general, the Q of a resonator involving a capacitor and an inductor can be determined from the Q values of the components, whether their losses come from series resistance or otherwise:

$$Q = \frac{1}{(1/Q_L + 1/Q_C)}$$

Mechanical Systems

For a single damped mass-spring system, the Q factor represents the effect of simplified viscous damping or drag, where the damping force or drag force is proportional to velocity. The formula for the Q factor is:

$$Q = \frac{\sqrt{Mk}}{D},$$

where M is the mass, k is the spring constant, and D is the damping coefficient, defined by the equation $F_{\text{damping}} = -Dv$, where v is the velocity.

Optical Systems

In optics, the Q factor of a resonant cavity is given by

$$Q = \frac{2\pi f_o \mathcal{E}}{P},$$

where f_o is the resonant frequency, ε is the stored energy in the cavity, and $P = -\frac{dE}{dt}$ is the power dissipated. The optical Q is equal to the ratio of the resonant frequency to the bandwidth of the cavity resonance.

The average lifetime of a resonant photon in the cavity is proportional to the cavity's Q. If the Q factor of a laser's cavity is abruptly changed from a low value to a high one, the laser will emit a pulse of light that is much more intense than the laser's normal continuous output. This technique is known as Q-switching.

Chapter 9

Equations of Motion

In mathematical physics, equations of motion are equations that describe the behaviour of a physical system in terms of its motion as a function of time. More specifically, the equations of motion describe the behaviour of a physical system as a set of mathematical functions in terms of dynamic variables: normally spatial coordinates and time are used, but others are also possible, such as momentum components and time.

The most general choice are generalised coordinates which can be any convenient variables characteristic of the physical system. The functions are defined in a Euclidean space in classical mechanics, but are replaced by curved spaces in relativity. If the dynamics of a system is known, the equations are the solutions to the differential equations describing the motion of the dynamics.

There are two main descriptions of motion: dynamics and kinematics. Dynamics is general, since momenta, forces and energy of the particles are taken into account.

In this instance, sometimes the term refers to the differential equations that the system satisfies (e.g., Newton's second law or Euler–Lagrange equations), and sometimes to the solutions to those equations.

However, kinematics is simpler as it concerns only spatial and time-related variables. In circumstances of constant acceleration, these simpler equations of motion are usually referred to as the "SUVAT" equations, arising from the definitions of kinematic quantities: displacement (S), initial velocity (U), final velocity (V), acceleration (A), and time (T).

Equations of motion can therefore be grouped under these main classifiers of motion. In all cases, the main *types* of motion are translations, rotations, oscillations, or any combinations of these. Historically, equations of motion initiated in classical mechanics and the extension to celestial mechanics, to describe the motion of massive objects.

Later they appeared in electrodynamics, when describing the motion of charged particles in electric and magnetic fields. With the advent of general relativity, the classical equations of motion became modified.

In all these cases the differential equations were in terms of a function describing the particle's trajectory in terms of space and time coordinates, as influenced by forces or energy transformations.

However, the equations of quantum mechanics can also be considered equations of motion, since they are differential equations of the wavefunction, which describes how a quantum state behaves analogously using the space and time coordinates of the particles.

There are analogs of equations of motion in other areas of physics, notably waves. These equations are explained below.

Qualitative

Equations of motion generally involve:

- a differential equation of motion, usually identified as some physical law and applying definitions of physical quantities, is used to set up an equation for the problem,
- setting the boundary and initial value conditions,
- a function of the position (or momentum) and time variables, describing the dynamics of the system,
- solving the resulting differential equation subject to the boundary and initial conditions.

The differential equation is a general description of the application and may be adjusted appropriately for a specific situation, the solution describes exactly how the system will behave for all times after the initial conditions, and according to the boundary conditions.

Quantitative

In Newtonian mechanics, an equation of motion M takes the general form of a second order ordinary differential equation (ODE) in the position r of the object:

$$M\left[r(t),\dot{r}(t),\ddot{r}(t),t\right]=0$$

where t is time, and each overdot denotes a time derivative.

The initial conditions are given by the *constant* values at $t = 0$:

$$r(0),\quad \dot{r}(0).$$

Another dynamical variable is the momentum p of the object, which can be used instead of r (though less commonly), i.e. a second order ODE in p:

$$\tilde{M}\left[p(t),\dot{p}(t),\ddot{p}(t),t\right]=0$$

with initial conditions (again *constant* values)

$$p(0),\quad \dot{p}(0).$$

The solution r (or p) to the equation of motion, combined with the initial values, describes of the system for all times after $t = 0$. For more than one particle, there are separate equations for each (this is contrary to a statistical ensemble of many particles in statistical mechanics, and a many-particle system in quantum mechanics - where all particles are described by a single probability distribution).

Sometimes, the equation will be linear and can be solved exactly. However in general, the equation is non-linear, and may lead to chaotic behaviour depending on how *sensitive* the system is to the initial conditions.

In the generalised Lagrangian mechanics, the generalised coordinates q (or generalised momenta p) replace the ordinary position (or momentum). Hamiltonian mechanics is slightly different, there are two first order equations in the generalised coordinates and momenta:

$$M\left[q(t),\dot{q}(t),t\right]=0,\quad \tilde{M}\left[p(t),\dot{p}(t),t\right]=0$$

where q is a tuple of generalised coordinates and similarly p is the tuple of generalised momenta. The initial conditions are similarly defined.

Kinematic Equations for one Particle

Kinematic Quantities

From the instantaneous position $r = r\ (t)$ (instantaneous meaning at an instant value of time t), the instantaneous velocity $v = v\ (t)$ and acceleration $a = a(t)$ have the general, coordinate-independent definitions;

$$v = \frac{dr}{dt}, \quad a = \frac{dv}{dt} = \frac{d^2r}{dt^2}$$

The rotational analogues are the angular position (angle the particle rotates about some axis) $\theta = \theta(t)$, angular velocity $\omega = \omega(t)$, and angular acceleration $a = a(t)$:

$$\omega = \hat{n}\frac{d\theta}{dt}, \quad \alpha = \frac{d\omega}{dt} = \hat{n}\frac{d^2\theta}{dt^2}$$

where

$$\hat{n} = \hat{e}_r \times \hat{e}_\theta$$

is a unit axial vector, pointing parallel to the axis of rotation, $\hat{e}_r$ = unit vector in direction of r, $\hat{e}_\theta$ = unit vector tangential to the angle.

NB: In these rotational definitions, the angle can be any angle about the specified axis of rotation. It is customary to use θ, but this does not have to be the polar angle used in polar coordinate systems.For a rotating rigid body, the following relations useful for describing the motion hold:

$$v = \omega \times r$$

$$a = \alpha \times r + \omega \times v$$

where r is a radial position.

Uniform Acceleration

Constant Linear Acceleration: These equations apply to a particle moving linearly, in three dimensions in a straight line, with constant acceleration. Since the vectors are collinear (parallel, and lie on the same line) - only the magnitudes of the vectors are necessary, hence non-bold letters are used for magnitudes, and because the motion is along a straight line, the problem effectively reduces from three dimensions to one.

Derivation:

$$v = at + v_0 \quad [1]$$

$$r = \frac{at^2}{2} + v_0 t + r_0 \quad [2]$$

$$r = r_0 + \left(\frac{v + v_0}{2}\right)t \quad [3]$$

$$v^2 = v_0^2 + 2a\left(r - r_0\right) \quad [4]$$

$$r = r_0 + vt - \frac{at^2}{2} \quad [5]$$

where r_0 and v_0 are the particle's initial position and velocity, r, v, a are the final position (displacement), velocity and acceleration of the particle after the time interval.

Here a is *constant* acceleration, or in the case of bodies moving under the influence of gravity, the standard gravity g is used. Note that each of the equations contains four of the five variables, so in this situation it is sufficient to know three out of the five variables to calculate the remaining two.

SUVAT Equations

In elementary physics the above formulae are frequently written:

$$v = at + u \quad [1]$$

$$s = ut + \frac{at^2}{2} \quad [2]$$

$$s = \left(\frac{v + u}{2}\right)t \quad [3]$$

$$v^2 = u^2 + 2as \quad [4]$$

$$s = vt - \frac{at^2}{2} \quad [5]$$

where u has replaced v_0, s replaces r, and $s_0 = 0$. They are often referred to as the "SUVAT" equations, eponymous from to the variables: s = displacement (s_0 = initial displacement), u = initial velocity, v = final velocity, a = acceleration, t = time.

Applications: Elementary and frequent examples in kinematics involve projectiles, for example a ball thrown upwards into the air. Given initial speed u, one can calculate how high the ball will travel before it begins to fall. The acceleration is local acceleration of gravity g. At this point one must remember that while these quantities appear to be scalars, the direction of displacement, speed and acceleration is important.

They could in fact be considered as uni-directional vectors. Choosing s to measure up from the ground, the acceleration a must be in fact "g, since the force of gravity acts downwards and therefore also the acceleration on the ball due to it.

At the highest point, the ball will be at rest: therefore $v = 0$. Using equation in the set above, we have:

$$s = \frac{v^2 - u^2}{-2g}.$$

Substituting and cancelling minus signs gives:

$$s = \frac{u^2}{2g}.$$

Constant Circular Acceleration

The analogues of the above equations can be written for rotation. Again these axial vectors must all be parallel (to the axis of rotation), so only the magnitudes of the vectors are necessary:

$$\begin{aligned}
\omega &= \omega_0 + \alpha t \\
\theta &= \theta_0 + \omega_0 t + \tfrac{1}{2}\alpha t^2 \\
\theta &= \theta_0 + \tfrac{1}{2}(\omega_0 + \omega)t \\
\omega^2 &= \omega_0^2 + 2\alpha(\theta - \theta_0) \\
\theta &= \theta_0 + \omega t - \tfrac{1}{2}\alpha t^2
\end{aligned}$$

where α is the constant angular acceleration, ω is the angular velocity, ω_0 is the initial angular velocity, θ is the angle turned through (angular displacement), θ_0 is the initial angle, and t is the time taken to rotate from the initial state to the final state.

General Planar Motion

These are the kinematic equations for a particle traversing a path in a plane, described by position $r = r(t)$. They are actually no more than the time derivatives of the position vector in plane polar coordinates in the context of physical quantities (like angular velocity ω). The position, velocity and acceleration of the particle are respectively:

$$r = r(r,\theta,t) = r\hat{e}_r$$

$$v = \hat{e}_r \frac{dr}{dt} + r\omega\hat{e}_\theta$$

$$a = \left(\frac{d^2r}{dt^2} - r\omega^2\right)\hat{e}_r + \left(r\alpha + 2\omega\frac{dr}{dt}\right)\hat{e}_\theta$$

where $\hat{\mathbf{e}}_r, \hat{\mathbf{e}}_\theta$, are the polar unit vectors. Notice for a the components $(-r\omega^2)$ and $2\omega dr/dt$ are the centripetal and Coriolis accelerations respectively.

Special cases of motion described be these equations are summarised qualitatively in the table below. Two have already been discussed above, in the cases that either the radial components or the angular components are zero, and the non-zero component of motion describes uniform acceleration.

State of motion	Constant *r*	Linear *r*	Quadratic *r*	Non-linear *r*
Constant θ	Stationary	Uniform translation (constant translational velocity)	Uniform translational acceleration	Non-uniform translation
Linear θ	Uniform angular motion in a circle (constant angular velocity)	Uniform angular motion in a spiral, constant radial velocity	Angular motion in a spiral, constant radial acceleration	Angular motion in a spiral, varying radial acceleration
Quadratic θ	Uniform angular acceleration in a circle	Uniform angular acceleration in a spiral, constant radial velocity	Uniform angular acceleration in a spiral, constant radial acceleration	Uniform angular acceleration in a spiral, varying radial acceleration
Non-linear θ	Non-uniform angular acceleration in a circle	Non-uniform angular acceleration in a spiral, constant radial velocity	Non-uniform angular acceleration in a spiral, constant radial acceleration	Non-uniform angular acceleration in a spiral, varying radial acceleration

General 3d Motion

It is certainly possible to derive analogue equations for motion in 3d space, but the equations become more complicated and unwieldy. Using the spherical coordinates (*r*, θ, □) with corresponding unit vectors $\hat{\mathbf{e}}_r, \hat{\mathbf{e}}_\theta, \hat{\mathbf{e}}_\phi$, the position, velocity, and acceleration are respectively:

$$
\begin{aligned}
r &= r(t) = r\hat{e}_r \\
v &= v\hat{e}_r + r\frac{d\theta}{dt}\hat{e}_\theta + r\frac{d\phi}{dt}\sin\theta\hat{e}_\phi \\
a &= \left(a - r\left(\frac{d\theta}{dt}\right)^2 - r\left(\frac{d\phi}{dt}\right)^2 \sin^2\theta\right)\hat{e}_r \\
&+ \left(r\frac{d^2\theta}{dt^2} + 2v\frac{d\theta}{dt} - r\left(\frac{d\phi}{dt}\right)^2 \sin\theta\cos\theta\right)\hat{e}_\theta \\
&+ \left(r\frac{d^2\phi}{dt^2}\sin\theta + 2v\frac{d\phi}{dt}\sin\theta + 2r\frac{d\theta}{dt}\frac{d\phi}{dt}\cos\theta\right)\hat{e}_\phi
\end{aligned}
$$

In the case of a constant □ this reduces to the planar equations above.

Harmonic Motion of One Particle

Translation: The kinematic equation of motion for a simple harmonic oscillator (SHO), oscillating in one dimension (the $\pm x$ direction) in a straight line is:

$$\frac{\mathrm{d}^2 x}{\mathrm{d}t^2} = -\omega^2 x$$

where ω is the angular frequency of the oscillatory motion, related to the general frequency f and the time period T (time taken for one cycle of oscillation):

$$\omega = 2\pi f = 2\pi / T$$

Many systems approximately execute simple harmonic motion (SHM). The complex harmonic oscillator is a superposition of simple harmonic oscillators:

$$\frac{\mathrm{d}^2 x}{\mathrm{d}t^2} = -\sum_n \omega_n^2 x$$

It is possible for simple harmonic motions to occur in any direction:

$$\frac{d^2 r}{dt^2} = -\sum_n \omega_n^2 r_n$$

known as a multidimensional harmonic oscillator. In cartesian coordinates, each component of the position will be a superposition of sinusiodal SHM.

Rotation: The rotational analogue of SHM in a straight line is angular oscillation about an axle or fulcrum:

$$\frac{d^2\theta}{dt^2} = -\omega^2\theta$$

where ω is still the angular frequency of the oscillatory motion - though *not* the angular velocity which is the rate of change of θ.

This form can be identified (at least approximately) as libration. The complex analogue is again a superposition of simple harmonic oscillators:

$$\frac{d^2\theta}{dt^2} = -\sum_n \omega_n^2\theta$$

Dynamic Equations of Motion

Newtonian Mechanics

It may be simple to write down the equations of motion in vector form using Newton's laws of motion, but the components may vary in complicated ways with spatial coordinates and time, and solving them is not easy. Often there is an excess of variables to solve for the problem completely, so Newton's laws are not the most efficient method for generally finding and solving for the motion of a particle. In simple cases of rectangular geometry, the use of Cartesian coordinates works fine, but other coordinate systems can become dramatically complex.

Newton's Second Law for Translation

The first developed and most famous is Newton's second law of motion, there are several ways to write and use it, the most general is:

$$F = \frac{dp}{dt}$$

where p = p(t) is the momentum of the particle and F = F(t) is the resultant external force acting on the particle (not any force the particle exerts) - in each case at time t. The law is also written more famously as:

$$F = ma$$

since m is a constant in Newtonian mechanics. However the momentum form is preferable since this is readily generalised to more complex systems, generalises to special and general relativity, and since momentum is a conserved quantity; with deeper fundamental significance than the position vector or its time derivatives.

For a number of particles, the equation of motion for one particle i influenced by other particles is:

$$\frac{dp_i}{dt} = F_E + \sum_{i \neq j} F_{ij}$$

where p_i = momentum of particle i, F_{ij} = force on particle i by particle j, and F_E = resultant external force (due to any agent not part of system). Particle i does not exert a force on itself.

Newton's Second Law for Rotation

For rigid bodies, Newton's second law for rotation takes the same form as for translation:

$$\tau = \frac{dL}{dt}$$

where I is the moment of inertia tensor.

Analogous to force and acceleration:

$$\tau = I \cdot \alpha$$

Likewise, for a number of particles, the equation of motion for one particle i is:

$$\frac{dL_i}{dt} = \tau_E + \sum_{i \neq j} \tau_{ij}$$

where L_i = angular momentum of particle i, τ_{ij} = torque on particle i by particle j, and τ_E = resultant external torque (due to any agent not part of system). Particle i does not exert a torque on itself.

Applications

Some examples of Newton's law include describing the motion of a pendulum:

$$mg\sin\theta = m\frac{\mathrm{d}^2(\ell\theta)}{\mathrm{d}t^2} + 0 \Rightarrow g\sin\theta = \ell\frac{\mathrm{d}^2\theta}{\mathrm{d}t^2}$$

a damped, driven harmonic oscillator:

$$F_0\sin(\omega t) = m\left(\frac{\mathrm{d}^2x}{\mathrm{d}t^2} + 2\zeta\omega_0\frac{\mathrm{d}x}{\mathrm{d}t} + \omega_0^2 x\right),$$

or a ball thrown in the air, in air currents (such as wind) described by a vector field of resistive forces R = R(x, y, z, t):

$$-\frac{GmM}{|r|^2}\hat{e}_r + R = m\frac{d^2r}{dt^2} + 0 \Rightarrow -\frac{GM}{|r|^2}\hat{e}_r + A = \frac{d^2r}{dt^2}$$

where G = gravitational constant, M = mass of the Earth and A is the acceleration of the projectile due to the air currents at position r and time t. Newton's law of gravity has been used. The mass m of the ball cancels.

Eulerian Mechanics

Euler developed Euler's laws of motion, analogous to Newton's laws, for the motion of rigid bodies.

Newton–Euler Equations

The Newton–Euler equations combine Euler's equations into one.

Analytical Mechanics

More effective equations of motion than Newton's laws are below.

Intrinsic Motion

Using all three coordinates of 3d space is unnecessary if there are constraints on the system. Generalised coordinates q(t) = [$q_1(t)$, $q_2(t)$ ... $q_N(t)$], where N is the total number of degrees of freedom the system has, are *any* set of coordinates used to define the configuration of the system, in the form of arc lengths or angles.

They are a considerable simplification to describe motion since they take advantage of the intrinsic constraints that limit the system's motion - i.e. the number of coordinates is reduced to a minimum, rather than demanding rote algebra to describe the constraints *and* the motion using *all* three coordinates.

Corresponding to generalised coordinates are:

- their time derivatives, the *generalised velocities*: $\dot{q} = dq / dt$,
- conjugate *"generalised" momenta*: $p = \partial L / \partial \dot{q} = \partial S / \partial q$,

where

- the *Lagrangian* is a function of the configuration q, the rate of change of configuration dq/d*t*, and time *t*: $L = L\left[\mathbf{q}(t), \dot{\mathbf{q}}(t), t\right]$,
- the *Hamiltonian* is a function of the configuration q, motion p, and time *t*: $H = H\left[q(t), p(t), t\right]$, and
- *Hamilton's principal function*, also called the *classical action* is a functional of *L*: $S[q,t] = \int_{t_1}^{t_2} L(q, \dot{q}, t)dt$.

The Lagrangian or Hamiltonian function is set up for the system using the q and p variables, then these are inserted into the Euler-Lagrange or Hamilton's equations to obtain differential equations of the system. These are solved for the coordinates and momenta.

Generalised Classical Equations of Motion

As the system evolves, q traces a path through configuration space (only some are shown). The path taken by the system (red) has a stationary action (dS = 0) under small changes in the configuration of the system (dq).

Principle of Least Action

All classical equations of motion can be derived from this variational principle:

$$\delta S = 0$$

stating the path the system takes through the configuration space is the one with the least action.

Euler-Lagrange Equations

The Euler-Lagrange equations are:

$$\frac{d}{dt}\left(\frac{\partial L}{\partial \dot{q}}\right) = \frac{\partial L}{\partial q}$$

After substituting for the Lagrangian, evaluating the partial derivatives, and simplifying, a second order ODE in each q_i is obtained.

Hamilton's Equations

Hamilton's equations are:

$$\dot{p} = -\frac{\partial H}{\partial q} \quad \dot{q} = +\frac{\partial H}{\partial p}$$

Notice the equations are symmetric (remain in the same form) by making these interchanges *simultaneously*:

$$p \rightleftharpoons q, \quad H \to -H.$$

After substituting the Hamiltonian, evaluating the partial derivatives, and simplifying, two first order ODEs in q_i and p_i are obtained.

Hamilton–Jacobi Equation

Hamilton's formalism can be rewritten as:

$$H = -\frac{\partial S}{\partial t}$$

Although the equation has a simple form, it's actually a *non-linear* PDE, first order in $N + 1$ variables, rather than $2N$ such equations. Due to the action S, it can be used to identify conserved quantities for mechanical systems, even when the mechanical problem itself cannot be solved fully, because any differentiable symmetry of the action of a physical system has a corresponding conservation law, a theorem due to Emmy Noether.

Electrodynamics

In electrodynamics, the force on a charged particle of charge q is the Lorentz force:

$$F = q\left(E + v \times B\right)$$

Combining with Newton's second law gives a first order differential equation of motion, in terms of position of the particle:

$$m\frac{d^2 r}{dt^2} = q\left(E + \frac{dr}{dt} \times B\right)$$

or its momentum:

$$\frac{dp}{dt} = q\left(E + \frac{p \times B}{m}\right)$$

The same equation can be obtained using the Lagrangian (and applying Lagrange's equations above) for a charged particle of mass m and charge q:

$$L = \frac{m}{2}\dot{r}\cdot\dot{r} + qA\cdot\dot{r} - q\phi$$

where A and ϕ are the electromagnetic scalar and vector potential fields. The Lagrangian indicates an additional detail: the canonical momentum in Lagrangian mechanics is given by:

$$P = \frac{\partial L}{\partial \dot{r}} = m\dot{r} + qA$$

instead of just mv, implying the motion of a charged particle is fundamentally determined by the mass and charge of the particle. The Lagrangian expression was first used to derive the force equation.

Alternatively the Hamiltonian (and substituting into the equations):

$$H = \frac{(P - qA)^2}{2m} - q\phi$$

can derive the Lorentz force equation.

Geodesic Equation of Motion

The above equations are valid in flat spacetime. In curved space spacetime, things become mathematically more complicated since there is no straight line; this is generalised and replaced by a *geodesic* of the curved spacetime (the shortest length of curve between two points). For curved manifolds with a metric tensor g, the metric provides the notion of arc length, the differential arc length is given by:

$$ds = \sqrt{g_{\alpha\beta} dx^{\alpha} dx^{\beta}}$$

and the geodesic equation is a second-order differential equation in the coordinates, the general solution is a family of geodesics:

$$\frac{d^2x^{\mu}}{ds^2} = -\Gamma^{\mu}{}_{\alpha\beta}\frac{dx^{\alpha}}{ds}\frac{dx^{\beta}}{ds}$$

where $\Gamma^{\mu}{}_{\alpha\beta}$ is a Christoffel symbol of the second kind, which contains the metric (with respect to the coordinate system).

Given the mass-energy distribution provided by the stress–energy tensor $T^{\alpha\beta}$, the Einstein field equations are a set of non-linear second-order partial differential equations in the metric, and imply the curvature of space time is equivalent to a gravitational field. Mass falling in curved spacetime is equivalent to a mass falling in a gravitational field - because gravity is a fictitious force.

The *relative acceleration* of one geodesic to another in curved spacetime is given by the *geodesic deviation equation*:

$$\frac{D^2\xi^{\alpha}}{ds^2} = -R^{\alpha}{}_{\beta\gamma\delta}\frac{dx^{\alpha}}{ds}\xi^{\gamma}\frac{dx^{\delta}}{ds}$$

where $\xi^{\alpha} = (x_2)^{\alpha} \times (x_1)^{\alpha}$ is the separation vector between two geodesics, D/ds (*not* just d/ds) is the covariant derivative, and $R^{\alpha}{}_{\beta\gamma d}$ is the Riemann curvature tensor, containing the Christoffel symbols. In other words, the geodesic deviation equation is the equation of motion for masses in curved spacetime, analogous to the Lorentz force equation for charges in an electromagnetic field.

For flat spacetime, the metric is a constant tensor so the Christoffel symbols vanish, and the geodesic equation has the solutions of straight lines. This is also the limiting case when masses move according to Newton's law of gravity.

Analogues for Waves and Fields

Field equations: Equations that describe the spatial dependence and time evolution of fields are called *field equations*. These include

- the Navier–Stokes equations for the velocity field of a fluid,
- Maxwell's equations for the electromagnetic field,
- the Einstein field equation for gravitation (Newton's law of gravity is a special case for weak gravitational fields and low velocities of particles).

Wave equations: Equations of wave motion are called *wave equations*. The solutions to a wave equation give the time-evolution and spatial dependence of the amplitude. Boundary conditions determine if the solutions describe travelling waves or standing waves. From classical equations of motion and field equations; mechanical and electromagnetic wave equations can be derived. The general linear wave equation in 3d is:

$$\nabla^2 X = \frac{1}{v^2}\frac{\partial^2 X}{\partial^2 t}$$

where $X = X(\mathrm{r}, t)$ is any mechanical or electromagnetic field amplitude, say:

- the transverse or longitudinal displacement of a vibrating rod, wire, cable, membrane etc.,
- the fluctuating pressure of a medium, sound pressure,
- the electric fields E or D, or the magnetic fields B or H,
- the voltage V or current I in an alternating current circuit,

and v is the phase velocity. Non-linear equations model the dependence of phase velocity on amplitude, replacing v by $v(X)$. There are other wave equations for very specific applications, non-linear equations arise in different mathematical forms.

In quantum mechanics, the analogue of the equation of motion is the Schrφdinger equation:

$$\hat{H}\Psi = i\hbar\frac{\partial \Psi}{\partial t}$$

where $\hat{H}$ is the Hamiltonian operator (rather than a function as above), ψ is the wavefunction and $\hbar$ is the reduced Planck constant. Setting up the Hamiltonian and inserting it into the equation results in a differential equation, the solution is the wavefunction as a function of space and time. There are also relativistic wave equations used in quantum field theory.

Relative Velocity

The relative velocity $\vec{v}_{\mathrm{BA}}$ (also $\vec{v}_{\mathrm{B/A}}$ or $\vec{v}_{\mathrm{BrelA}}$) is the velocity of an object or observer B in the rest frame of another object or observer A. If it is constant,

$$\vec{v}_{\mathrm{BA}} = -\vec{v}_{\mathrm{AB}}$$

where $\vec{v}_{\mathrm{AB}}$ is A's velocity in the rest frame of B.

Classical Mechanics

In the Newtonian limit where approximately the Galilean transformation

$$\vec{r}' = \vec{r} - \vec{v}t$$
$$t' = t$$

holds, it is identical with the vector difference between the velocities of A and B, as evaluated in terms of a single coordinate system:

$$\vec{v}_{BA} = \vec{v}_B - \vec{v}_A \, .$$

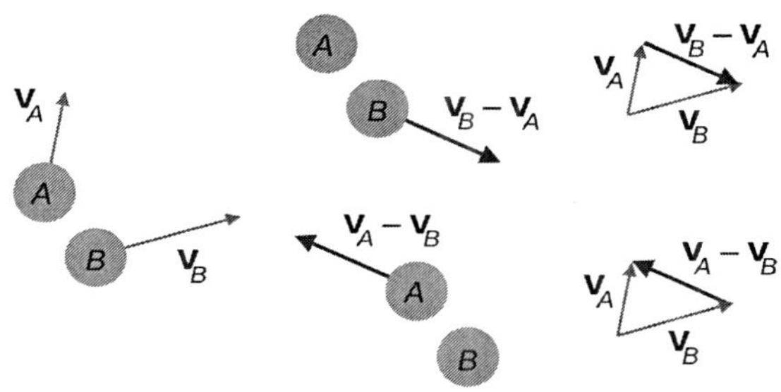

Figure: *Relative velocities between two particles in classical mechanics.*

Due to Einstein's Special Relativity (SR), doesn't hold in general. The expression "the velocity of A relative to B" is shorthand for "the velocity of A in the coordinate system where B is always at rest".

Special Relativity Theory

According to SR, the vacuum speed of light is isotropically equal to a universal constant c in any coordinate systems. Inter alia, it leads to the fact that

- apart from Newtonian limit, velocities are not additive quantities, and
- the difference velocity between A and B is not equal to their relative velocity and particularly has a smaller absolute value. Whereas the maximum difference speed between two objects is 2c, the maximum absolute value of a relative velocity is equal to c.

To get $\vec{v}_{BA}$ from $\vec{v}_A$ and $\vec{v}_B$ in an arbitrary reference frame, it's necessary to Lorentz transformation the latter into the rest frame of A. If $\vec{v}_A$ and $\vec{v}_B$ are collinear, the formula $v_{BA} = \dfrac{v_B - v_A}{1 - \dfrac{v_A v_B}{c^2}}$ holds.

Example

Joe and Sara are driving in the same direction. Joe's velocity is 90 km/h and Sara's 100 km/h. If we take Joe's velocity as $\vec{v}_B$ and Sara's $\vec{v}_A$ then

$$\vec{v}_{A \text{ rel } B} = 100 - 90 = 10 \text{ km/h}.$$

This is the velocity observed by Joe. Joe sees Sara moving at 10 km/h.

Polar Coordinates

In polar coordinates, a two-dimensional velocity is described by a radial velocity, defined as the component of velocity away from or toward the origin (also known as *velocity made good*), and an angular velocity, which is the rate of rotation about the origin (with positive quantities representing counter-clockwise rotation and negative quantities representing clockwise rotation, in a right-handed coordinate system).

The radial and angular velocities can be derived from the Cartesian velocity and displacement vectors by decomposing the velocity vector into radial and transverse components. The transverse velocity is the component of velocity along a circle centred at the origin.

$$v = v_T + v_R$$

where

v_T is the transverse velocity

v_R is the radial velocity.

The *magnitude of the radial velocity* is the dot product of the velocity vector and the unit vector in the direction of the displacement.

$$v_R = \frac{v \cdot r}{|r|}$$

where

r is displacement.

The *magnitude of the transverse velocity* is that of the cross product of the unit vector in the direction of the displacement and the velocity vector. It is also the product of the angular speed ω and the magnitude of the displacement.

$$v_T = \frac{|r \times v|}{|r|} = \omega |r|$$

Such that:

$$\omega = \frac{|r \times v|}{|r|^2}.$$

Angular momentum in scalar form is the mass times the distance to the origin times the transverse velocity, or equivalently, the mass times the distance squared times the angular speed. The sign convention for angular momentum is the same as that for angular velocity.

$$L = mrv_T = mr^2\omega$$

where

m is mass

$r = ||r||$.

The expression mr^2 is known as moment of inertia. If forces are in the radial direction only with an inverse square dependence, as in the case of a gravitational orbit, angular momentum is constant, and transverse speed is inversely proportional to the distance, angular speed is inversely proportional to the distance squared, and the rate at which area is swept out is constant. These relations are known as Kepler's laws of planetary motion.

Acceleration

In physics, acceleration is the rate at which the velocity of a body changes with time. In general, velocity and acceleration are vector quantities, with magnitude and direction, though in many cases only magnitude is considered (sometimes with negative values for deceleration). Acceleration is accompanied by a force, as described by Newton's Second Law; the force, as a vector, is the product of the mass of the object being accelerated and the acceleration (vector). The SI unit of acceleration is the meter per second squared (m/s^2).

For example, an object such as a car that starts from standstill, then travels in a straight line at increasing speed, is accelerating in the direction of travel. If the car changes direction at constant speedometer reading, there is strictly speaking an acceleration although it is often not so described; passengers in the car will experience a force pushing them back into their seats in linear acceleration, and a sideways force on changing direction. If the speed of the car decreases, it is usual and meaningful to speak of deceleration; mathematically it is acceleration in the opposite direction to that of motion.

Definition and Properties

Acceleration is the rate of change of velocity. At any point on a trajectory, the magnitude of the acceleration is given by the rate of change of velocity in both magnitude and direction at that point. The true acceleration at time t is found in the limit as time interval $\Delta t \to 0$ of $\Delta v/\Delta t$

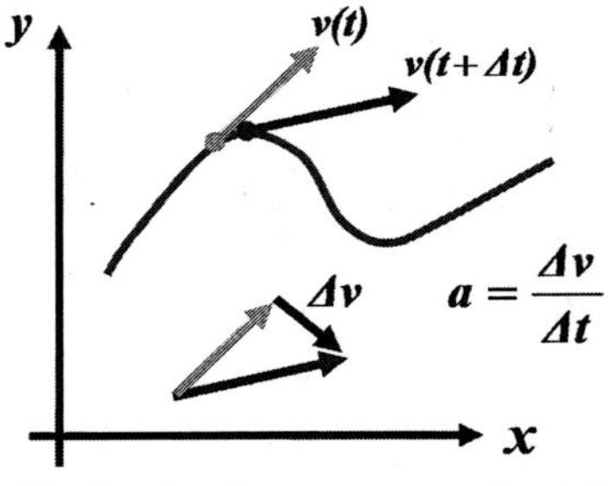

Mathematically, instantaneous acceleration—acceleration over an infinitesimal interval of time—is the rate of change of velocity over time:

$$a = \lim_{\Delta t \to 0} \frac{\Delta v}{\Delta t} = \frac{dv}{dt},$$

i.e., the derivative of the velocity vector as a function of time.

(Here and elsewhere, if motion is in a straight line, vector quantities can be substituted by scalars in the equations.)

Average acceleration over a period of time is the change in velocity (Δ/v) divided by the duration of the period (Δ/t)

$$\bar{a} = \frac{\Delta v}{\Delta t}.$$

Acceleration has the dimensions of velocity (L/T²) divided by time, i.e., L/T². The SI unit of acceleration is the metre per second per second (m/s²); this can be called more meaningfully "metre per second per second", as the velocity in metres per second changes by the acceleration value, every second.

An object moving in a circular motion—such as a satellite orbiting the earth—is accelerating due to the change of direction of motion, although the magnitude (speed) may be constant. When an object is executing such a motion where it changes direction, but not speed, it is said to be undergoing *centripetal* (directed towards the centre) acceleration. Oppositely, a change in the speed of an object, but not its direction of motion, is a *tangential* acceleration. Proper acceleration, the acceleration of a body relative to a free-fall condition, is measured by an instrument called an accelerometer.

In classical mechanics, for a body with constant mass, the (vector) acceleration of the body's centre of mass is proportional to the net force vector (i.e., sum of all forces) acting on it (Newton's second law):

$$F = ma \quad \rightarrow \quad a = F / m$$

where F is the net force acting on the body, m is the mass of the body, and a is the centre-of-mass acceleration. As speeds approach the speed of light, relativistic effects become increasingly large and acceleration becomes less.

Tangential and Centripetal Acceleration

The velocity of a particle moving on a curved path as a function of time can be written as:

$$v(t) = v(t)\frac{v(t)}{v(t)} = v(t)u_t(t),$$

with $v(t)$ equal to the speed of travel along the path, and

$$u_t = \frac{v(t)}{v(t)},$$

a unit vector tangent to the path pointing in the direction of motion at the chosen moment in time. Taking into account both the changing speed *v(t)* and the changing direction of u_t, the acceleration of a particle moving on a curved path can be written using the chain rule of differentiation for the product of two functions of time as:

$$\begin{aligned} a &= \frac{dv}{dt} \\ &= \frac{dv}{dt} u_t + v(t) \frac{du_t}{dt} \\ &= \frac{dv}{dt} u_t + \frac{v^2}{r} u_n , \end{aligned}$$

where u_n is the unit (inward) normal vector to the particle's trajectory (also called *the principal normal*), and r is its instantaneous radius of curvature based upon the osculating circle at time *t*. These components are called the tangential acceleration and the normal or radial acceleration (or centripetal acceleration in circular motion).

Geometrical analysis of three-dimensional space curves, which explains tangent, (principal) normal and binormal, is described by the Frenet–Serret formulas.

Special Cases

Uniform Acceleration: *Uniform* or *constant* acceleration is a type of motion in which the velocity of an object changes by an equal amount in every equal time period.

A frequently cited example of uniform acceleration is that of an object in free fall in a uniform gravitational field. The acceleration of a falling body in the absence of resistances to motion is dependent only on the gravitational field strength *g* (also called *acceleration due to gravity*). By Newton's Second Law the force, *F*, acting on a body is given by:

$$F = mg$$

Due to the simple algebraic properties of constant acceleration in the one-dimensional case (that is, the case of acceleration aligned with the initial velocity), there are simple formulas that relate the following quantities: displacement, initial velocity *u*, final velocity *v*, acceleration *a*, and time *t*:

$$v = u + at$$

$$s = ut + \frac{1}{2}at^2 = \frac{(u+v)t}{2}$$

$$|v|^2 = |u|^2 + 2a \cdot s$$

where

s = displacement

u = initial velocity

v = final velocity

a = uniform acceleration

t = time.

In the case of uniform acceleration of an object that is initially moving in a direction not aligned with the acceleration, the motion can be resolved into two orthogonal parts, one of constant velocity and the other according to the above equations. As Galileo showed, the net result is parabolic motion, as in the trajectory of a cannonball, neglecting air resistance.

Circular Motion

Uniform circular motion, that is constant speed along a circular path, is an example of a body experiencing acceleration resulting in velocity of a constant magnitude but change of direction. In this case, because the direction of the object's motion is constantly changing, being tangential to the circle, the object's linear velocity vector also changes, but its speed does not. This acceleration is a radial acceleration since it is always directed toward the centre of the circle and takes the magnitude:

$$\mathrm{a} = \frac{v^2}{r}$$

where v is the object's linear speed along the circular path. Equivalently, the radial acceleration vector (a) may be calculated from the object's angular velocity ω, whence:

$$a = -\omega^2 r$$

where r is a vector directed from the centre of the circle and equal in magnitude to the radius. The negative shows that the acceleration vector is directed towards the centre of the circle (opposite to the radius).

The acceleration, hence also the force acting on a body in uniform circular motion, is directed *toward* the centre of the circle; that is, it is centripetal. Whereas the so-called 'centrifugal force' appearing to act outward on the body is really a pseudo force experienced in the frame of reference of the body in circular motion, due to the body's linear momentum at a tangent to the circle.

With nonuniform circular motion, i.e., the speed along the curved path changes, a transverse accleration is produced equal to the rate of change of the angular speed around the circle times the radius of the circle. That is,

$$a = r\alpha.$$

The transverse (or tangential) acceleration is directed at right angles to the radius vector and takes the sign of the angular acceleration (α).

Chapter 10

Principles of Number Theorems

Erdõs–Graham Problem

In combinatorial number theory, the Erdõs–Graham problem is the problem of proving that, if the set {2, 3, 4, ...} of integers greater than one is partitioned into finitely many subsets, then one of the subsets can be used to form an Egyptian fraction representation of unity. That is, for every $r > 0$, and every r-colouring of the integers greater than one, there is a finite monochromatic subset S of these integers such that

$$\sum_{n \in S} \frac{1}{n} = 1.$$

In more detail, Paul Erdõs and Ronald Graham conjectured that, for sufficiently large r, the largest member of S could be bounded by b^r for some constant b independent of r. It was known that, for this to be true, b must be at least e.

Ernie Croot proved the conjecture as part of his Ph.D thesis, and later (while a post-doctoral student at UC Berkeley) published the proof in the *Annals of Mathematics*. The value Croot gives for b is very large: it is at most e^{167000}.

Croot's result follows as a corollary of a more general theorem stating the existence of Egyptian fraction representations of unity for sets C of smooth numbers in intervals of the form $[X, X^{1+\delta}]$, where C contains sufficiently many numbers so that the sum of their reciprocals is at least six. The Erdõs–Graham conjecture follows from this result by showing that one can find an interval of this form in which the sum of the reciprocals of all smooth numbers is at least $6r$;

therefore, if the integers are r-coloured there must be a monochromatic subset C satisfying the conditions of Croot's theorem.

Erdõs–Fuchs Theorem

In mathematics, in the area of combinatorial number theory, the Erdõs–Fuchs theorem is a statement about the number of ways that numbers can be represented as a sum of two elements of a given set, stating that the average order of this number cannot be close to being a linear function.

The theorem is named after Paul Erdõs and Wolfgang Heinrich Johannes Fuchs.

Statement

Let A be a subset of the natural numbers and $r(n)$ denote the number of ways that a natural number n can be expressed as the sum of two elements of A (taking order into account). We consider the average

$$R(n) = \frac{r(1) + r(2) + \cdots + r(n)}{n}.$$

The theorem states that

$$R(n) = C + O\left(n^{-3/4-\varepsilon}\right)$$

cannot hold unless $C = 0$.

Erdõs–Szemerédi Theorem

In arithmetic combinatorics, the Erdõs–Szemerédi theorem, proven by Paul Erdõs and Endre Szemerédi in 1983, asserts the existence of positive constants c, such that

$$\max(|A+A|, |A \cdot A|) \geq c|A|^{1+\varepsilon}$$

whenever A is a finite non-empty set of real numbers of cardinality $|A|$, $A+A = \{a+b : a,b \in A\}$ where is the sum-set of A with itself, and $A \cdot A = \{ab : a,b \in A\}$. Note that it is possible for $A+A$ to be of comparable size to A if A is an arithmetic progression, and it is possible for $A \cdot A$ to be of comparable size to A if A is a geometric progression. The Erdõs–Szemerédi theorem can thus be viewed as an assertion that it is not possible for a large set to behave like an arithmetic progression and as a geometric progression simultaneously. It can also be viewed as an assertion that the real line does not contain any set resembling a finite subring or finite subfield; it is the first example of what is now known as the *sum-product phenomenon*, which is now known to hold in a wide variety of rings and fields, including finite fields.

It was conjectured by Erdõs and Szemerédi that one can take arbitrarily close to 1. The best result in this direction currently is by Solymosi, who showed that one can take arbitrarily close to 1/3.

Euler's Theorem

In number theory, Euler's theorem (also known as the Fermat–Euler theorem or Euler's totient theorem) states that if n and a are coprime positive integers, then

$$a^{\varphi(n)} \equiv 1 \pmod{n}$$

where $\varphi(n)$ is Euler's totient function. (The notation is explained in the article Modular arithmetic.) It was first stated and proved by Leonhard Euler in 1736.

There is a converse of Euler's theorem: if the above congruence is true then a and n must be coprime.

The theorem is a generalization of Fermat's little theorem, and is further generalised by Carmichael's theorem.

The theorem may be used to easily reduce large powers modulo n. For example, consider finding the ones place decimal digit of 7^{222}, i.e. 7^{222} (mod 10). Note that 7 and 10 are coprime, and $\varphi(10) = 4$. So Euler's theorem yields 7^4 a– 1 (mod 10), and we get 7^{222} a– $7^{4\times55+2}$ a– $(7^4)^{55}\times7^2$ a– $1^{55}\times7^2$ a– 49 a– 9 (mod 10).

In general, when reducing a power of a modulo n (where a and n are coprime), one needs to work modulo $\varphi(n)$ in the exponent of a:

if x a– y (mod $\varphi(n)$), then a^x a– a^y (mod n).

Euler's theorem also forms the basis of the RSA encryption system: encryption and decryption in this system together amount to exponentiating the original text by $k\varphi(n)+1$ for some positive integer k, so Euler's theorem shows that the decrypted result is the same as the original.

Faltings' Theorem

In number theory, the Mordell conjecture is the conjecture made by Mordell (1922) that a curve of genus greater than 1 over the field Q of rational numbers has only finitely many rational points. The conjecture was later generalised by replacing Q by a finite extension. It was proved by Gerd Faltings (1983), and is now known as Faltings' theorem.

Background

Let C be a non-singular algebraic curve of genus g over Q. Then the set of rational points on C may be determined as follows:

- Case $g = 0$: no points or infinitely many; C is handled as a conic section.
- Case $g = 1$: no points, or C is an elliptic curve and its rational points form a finitely generated abelian group (*Mordell's Theorem*, later generalised to the Mordell–Weil theorem). Moreover Mazur's torsion theorem restricts the structure of the torsion subgroup.
- Case $g > 1$: according to the Mordell conjecture, now Faltings' Theorem, C has only a finite number of rational points.

Proofs

Faltings' original proof used the known reduction to a case of the Tate conjecture, and a number of tools from algebraic geometry, including the theory of Néron models. A very different proof, based on diophantine approximation, was found by Paul Vojta. A more elementary variant of Vojta's proof was given by Enrico Bombieri.

Consequences

Faltings' 1983 paper had as consequences a number of statements which had previously been conjectured:

- The Mordell conjecture that a curve of genus greater than 1 over a number field has only finitely many rational points;
- The Shafarevich conjecture that there are only finitely many isomorphism classes of abelian varieties of fixed dimension and fixed polarisation degree over a fixed

number field with good reduction outside a given finite set of places; and

- The *Isogeny theorem* that abelian varieties with isomorphic Tate modules (as Q_l-modules with Galois action) are isogenous.

The reduction of the Mordell conjecture to the Shafarevich conjecture was due to Parshin (1971). A sample application of Faltings' theorem is to a weak form of Fermat's Last Theorem: for any fixed $n > 4$ there are at most finitely many primitive integer solutins to $a^n + b^n = c^n$, since for such n the curve $x^n + y^n = 1$ has genus greater than 1.

Generalisations

Because of the Mordell–Weil theorem, Faltings' theorem can be reformulated as a statement about the intersection of a curve C with a finitely generated subgroup Γ of an abelian variety A. Generalising by replacing C by an arbitrary subvariety of A and Γ by an arbitrary finite-rank subgroup of A leads to the Mordell–Lang conjecture, which has been proved.

Another higher-dimensional generalisation of Faltings' theorem is the Bombieri–Lang conjecture that if X is a pseudo-canonical variety (i.e., variety of general type) over a number field k, then $X(k)$ is not Zariski dense in X. Even more general conjectures have been put forth by Paul Vojta.

The Mordell conjecture for function fields was proved by Manin (1963) and by Grauert (1965). Coleman (1990) found and fixed a gap in Manin's proof.

Fermat Polygonal Number Theorem

In additive number theory, the Fermat polygonal number theorem states that every positive integer is a sum of at most n n-gonal numbers.

That is, every positive number can be written as the sum of three or fewer triangular numbers, and as the sum of four or fewer square numbers, and as the sum of five or fewer pentagonal numbers, and so on. Three such representations of the number 17, for example, are shown below:

17 = 10 + 6 + 1 (*triangular numbers*)

17 = 16 + 1 (*square numbers*)

17 = 12 + 5 (*pentagonal numbers*).

The theorem is named after Pierre de Fermat, who stated it without proof, promising to write it in a sepaate work that never appeared. Joseph Louis Lagrange proved the square case in 1770, which states that every positive number can be represented as a sum of four squares, for example, 7 = 4 + 1 + 1 + 1.. Gauss proved the triangular case in 1796, commemorating the occasion by writing in his diary the line "ÅÕÑÇÊÁ! num = Δ + Δ + ≤, and published a proof in his book Disquisitiones Arithmeticae. For this reason, Gauss' result is sometimes known as the Eureka theorem. The full polygonal number theorem was not resolved until it was finally proven by Cauchy in 1813. The proof of Nathanson (1987) is based on the following lemma due to Cauchy:

For odd positive integers a and b such that $b^2 < 4a$ and $3a < b^2 + 2b + 4$ we can find nonnegative integers s, t, u, and v such that $a = s^2 + t^2 + u^2 + v^2$ and $b = s + t + u + v$.

Fermat's Theorem on Sums of Two Squares

In additive number theory, Pierre de Fermat's theorem on sums of two squares states that an odd prime p is expressible as

$$p = x^2 + y^2,$$

with x and y integers, if and only if

$$p \equiv 1 \pmod 4.$$

For example, the primes 5, 13, 17, 29, 37 and 41 are all congruent to 1 modulo 4, and they can be expressed as sums of two squares in the following ways:

$$5 = 1^2 + 2^2, \quad 13 = 2^2 + 3^2, \quad 17 = 1^2 + 4^2, \quad 29 = 2^2 + 5^2, \quad 37 = 1^2 + 6^2, \quad 41 = 4^2 + 5^2.$$

On the other hand, the primes 3, 7, 11, 19, 23 and 31 are all congruent to 3 modulo 4, and none of them can be expressed as the sum of two squares.

Albert Girard was the first to make the observation (in 1632) and Fermat was first to claim a proof of it. Fermat announced this theorem in a letter to Marin Mersenne dated

December 25, 1640; for this reason this theorem is sometimes called *Fermat's Christmas Theorem.*

Since the Brahmagupta–Fibonacci identity implies that the product of two integers that can be written as the sum of two squares is itself expressible as the sum of two squares, by applying Fermat's heorem to the prime factorisation of any positive integer n, we see that if all of n's odd prime factors congruent to 3 modulo 4 occur to an even exponent, it is expressible as a sum of two squares. The converse also holds.

Proofs of Fermat's Theorem on Sums of Two Squares

Fermat usually did not write down proofs of his claims, and he did not provide a proof of this statement. The first proof was found by Euler after much effort and is based on infinite descent. He announced it in a letter to Goldbach on April 12, 1749. Lagrange gave a proof in 1775 that was based on his study of quadratic forms. This proof was simplified by Gauss in his *Disquisitiones Arithmeticae* (art. 182). Dedekind gave at least two proofs based on the arithmetic of the Gaussian integers.

There is an elegant proof using Minkowski's theorem about convex sets. Simplifying an earlier short proof due to Heath-Brown (who was inspired by Liouville's idea), Zagier presented a one-sentence proof of Fermat's assertion.

Related Results

Fermat announced two related results fourteen years later. In a letter to Blaise Pascal dated September 25, 1654 he announced the following two results for odd primes :

- $p = x^2 + 2y^2 \Leftrightarrow p \equiv 1 \text{ or } p \equiv 3 \pmod{8}$,
- $p = x^2 + 3y^2 \Leftrightarrow p \equiv 1 \pmod{3}$.

He also wrote:

If two primes which end in 3 or 7 and surpass by 3 a multiple of 4 are multiplied, then their product will be composed of asquare and the quintuple of another square.

In other words, if p, q are of the form $20k + 3$ or $20k + 7$, then $pq = x^2 + 5y^2$. Euler later extended this to the conjecture that

- $p = x^2 + 5y^2 \Leftrightarrow p \equiv 1 \text{ or } p \equiv 9 \pmod{20}$,

- $2p = x^2 + 5y^2 \Leftrightarrow p \equiv 3 \text{ or } p \equiv 7 \pmod{20}$.

  Both Fermat's assertion and Euler's conjecture were established by Lagrange.

Freiman's Theorem

In mathematics, Freiman's theorem is a combinatorial result in number theory. In a sense it accounts for the approximate structure of sets of integers that contain a high proportion of their internal sums, taken two at a time.

The formal statement is:

Let A be a finite set of integers such that the sumset

$$A + A$$

is small, in the sense that

$$|A + A| < c|A|$$

for some constant . There exists an n-dimensional arithmetic progression of length

$$c'|A|$$

that contains A, and such that c' and n depend only on c.

A simple instructive case is the following. We always have $|A+A|$ e– $2|A|$-1, with equality precisely when A is an arithmetic progression.

This result is due to Gregory Freiman (1966). Much interest in it, and applications, stemmed from a new proof by Imre Z. Ruzsa (1994).

Fundamental Lemma (Langlands Programme)

In the theory of automorphic forms, an area of mathematics, the fundamental lemma relates orbital integrals on a reductive group over a local field to stable orbital integrals on its endoscopic groups. It was conjectured by Langlands (1983) in the course of developing the Langlands programme.

The fundamental lemma was proved by Gérard Laumon and Ngô B£o Chβu in the case of unitary groups and then by Ngô for general reductive groups, building on a series of important reductions made by Jean-Loup Waldspurger to the case of Lie algebras. Time magazine placed Ngô's proof on the list of the "Top 10 scientific discoveries of 2009. In 2010 Ngô was awarded the Fields medal for this proof.

Motivation and History

Robert Langlands outlined a strategy for proving local and global Langlands conjectures using the Arthur–Selberg trace formula, but in order for this approach to work, the geometric sides of the trace formula for different groups must be related in a particular way. This relationship takes the form of identities between orbital integrals on reductive groups G and H over a nonarchimedean local field F, where the group H, called an endoscopic group of G, is constructed from G and some additional data.

The first case considered was $G = SL_2$ (Labesse & Langlands 1979). Langlands and Shelstad (1987) then developed the general framework for the theory of endoscopic transfer and formulated specific conjectures. However, during the next two decades only partial progress was made towards proving the fundamental lemma. Harris called it a "bottleneck limiting progress on a host of arithmetic questions". Langlands himself, writing on the origins of endoscopy, commented:

> *... it is not the fundamental lemma as such that is critical for the analytic theory of automorphic forms and for the arithmetic of Shimura varieties; it is the stabilised (or stable) trace formula, the reduction of the trace formula itself to the stable trace formula for a group and its endoscopic groups, and the stbilisation of the Grothendieck–Lefschetz formula. None of these are possible without the fundamental lemma and its absence rendered progress almost impossible for more than twenty years.*

Statement

The fundamental lemma states that an orbital integral O for a group G is equal to a stable orbital integral SO for an endoscopic group H, up to a transfer factor Δ (Nadler 2012):

$$SO_{\gamma_H}(1_{K_H}) = \Delta(\gamma_H, \gamma_G) O^{\kappa}_{\gamma_G}(1_{K_G})$$

where

- F is a local field
- G is an unramified group defined over F, in other words a quasi-split reductive group defined over F that splits over an unramified extension of F
- H is an unramified endoscopic group of G associated to κ
- K_G and K_H are hyperspecial maximal compact subgroups of G and H, which means roughly that they are the subgroups of points with coefficients in the ring of integers of F.
- 1_{KG} and 1_{KH} are the characteristic functions of K_G and K_H.
- $\Delta(\tilde{a}_H, \tilde{a}_G)$ is a transfer factor, a certain elementary expression depending on γ_H and γ_G
- γ_H and γ_G are elements of G and H representing stable conjugacy classes, such that the stable conjugacy class of G is the transfer of the stable conjugacy class of H.
- κ is a character of the group of conjugacy classes in the stable conjugacy class of γ_G
- SO and O are stable orbital integrals and orbital integrals depending on their parametres.

Approaches

Shelstad (1982) proved the fundamental lemma for archimedian fields.

Waldspurger (1991) verified the fundamental lemma for general linear groups.

Kottwitz (1992) and Blasiu & Rogawski (1992) verified some cases of the fundamental lemma for 3-dimensional unitary groups.

Hales (1997) and Weissauer (2009) verified the fundamental lemma for the symplectic and general symplectic groups Sp_4, GSp_4.

A paper of George Lusztig and David Kazhdan pointed out that orbital integrals could be interpreted as counting points on certain algebraic varieties over finite fields.

Further, the integrals in question can be computed in a way that depends only on the residue field of F; and the issue can be reduced to the Lie algebra version of the orbital integrals.

Then the problem was restated in terms of the Springer fibre of algebraic groups.

The circle of ideas was connected to a purity conjecture; Laumon gave a conditional proof based on such a conjecture, for unitary groups.

Laumon and Ngô (2008) then proved the fundamental lemma for unitary groups, using Hitchin fibration introduced by Ngô (2006), which is an abstract geometric analogue of the Hitchin system of complex algebraic geometry. Waldspurger (2006) showed for Lie algebras that the function field case implies the fundamental lemma over all local fields, and Waldspurger (2008) showed that the fundamental lemma for Lie algebras implies the fundamental lemma for groups. .

Fundamental Lemma of Sieve Theory

In number theory, the fundamental lemma of sieve theory is any of several results that systematise the process of applying sieve methods to particular problems. Halberstam & Richert write:

A curious feature of sieve literature is that while there is frequent use of Brun's *method* there are only a few attempts to formulate a general Brun *theorem* (such as Theorem 2.1); as a result there are surprisingly many papers which repeat in considerable detail the steps of Brun's argument.

Diamond & Halberstam attribute the terminology *Fundamental Lemma* to Jonas Kubilius.

Common Notation

We use these notations:

- A is a set of X positive integers, and A_d is its subset of integers divisible by d
- $w(d)$ and R_d are functions of A and of d that estimate the number of elements of A that are divisible by d, according to the formula

$$|A_d| = \frac{w(d)}{d} X + R_d .$$

Thus $w(d) / d$ represents an approximate density of members divisible by d, and R_d represents an error or remainder term.

- P is a set of primes, and $P(z)$ is its the product of those primes $\leq z$
- $S(A, P, z)$ is the number of elements of A not divisible by any prime in P that is $\leq z$
- κ is a constant, called the sifting density, that appears in the assumptions below. It is a weighted average of the number of residue classes sieved out by each prime.

Fundamental Lema of the Combinatorial Sieve

This formulation is from Tenenbaum. Other formulations are in Halberstam & Richert, in Greaves, and in Friedlander & Iwaniec. We make the assumptions:

- $w(d)$ is a multiplicative function.
- The sifting density κ satisfies, for some constant C and any real numbers η and ξ with $2 \leq \eta \leq \xi$:

$$\prod_{\eta \leq p \leq \xi} \left(1 - \frac{w(p)}{p}\right)^{-1} < \left(\frac{\ln \xi}{\ln \eta}\right)^{\kappa} \left(1 + \frac{C}{\ln \eta}\right).$$

There is a parametre u e– 1 that is at our disposal. We have uniformly in A, X, z, and u that

$$S(a, P, z) = X \prod_{p \leq z, p \in P} \left(1 - \frac{w(p)}{p}\right) \{1 + O(u^{-u/2})\} + O\left(\sum_{d \leq z^u, d | P(z)} |R_d|\right).$$

In applications we pick u to get the best error term. In the sieve it represents the number of levels of the inclusion–exclusion principle.

Fundamental Lemma of the Selberg Sieve

This formulation is from Halberstam & Richert. Another formulation is in Diamond & Halberstam.

We make the assumptions:

- $w(d)$ is a multiplicative function.
- The sifting density κ satisfies, for some constant C and any real numbers η and ξ with $2 \leq \eta \leq \xi$:

$$\sum_{\eta \leq p \leq \xi} \frac{w(p) \ln p}{p} < \kappa \ln \frac{\xi}{\eta} + C.$$

- $w(p) / p < 1 - c$ for some small fixed c and all p

- $|R_d| \le \omega(d)$ where $\omega(d)$ is the number of distinct prime divisors of d.

The fundamental lemma has almost the same form as for the combinatorial sieve. Write $u = \ln X / \ln z$. The conclusion is:

$$S(a,P,z) = X \prod_{p \le z,\, p \in P} \left(1 - \frac{w(p)}{p}\right)\{1 + O(e^{-u/2})\}.$$

Note that u is no longer an independent parametre at our disposal, but is controlled by the choice of z.

Note that the error term here is weaker than for the fundamental lemma of the combinatorial sieve. Halberstam & Richert remark: "Thus it is not true to say, as has been asserted from time to time in the literature, that Selberg's sieve is always better than Brun's."

Gelfond–Schneider Theorem

In mathematics, the Gelfond–Schneider theorem establishes the transcendence of a large class of numbers. It was originally proved independently in 1934 by Aleksandr Gelfond and Theodor Schneider. The Gelfond–Schneider theorem answers affirmatively Hilbert's seventh problem.

Statement

If α and β are algebraic numbers with $\alpha \neq 0,1$ and if β is not a rational number, then any value of $\alpha^\beta = \exp(\beta \log \alpha)$ is a transcendental number.

Comments

- The values of α and β are not restricted to real numbers; complex numbers are allowed (and never rational, even if both the real and imaginary parts are rational).
- In general, $\alpha^\beta = \exp(\beta \log \alpha)$ is multivalued, where "log" stands for the complex logarithm. This accounts for the phrase "any value of" in the theorem's statement.
- An equivalent formulation of the theorem is the following: if α and γ are nonzero algebraic numbers, and we take any non-zero logarithm of α, then $(\log \gamma)/(\log \alpha)$ is either rational or transcendental. This may be expressed as saying that if $\log \alpha$, $\log \gamma$ are linearly independent over the rationals, then they are linearly

independent over the algebraic numbers. The generalisation of this statement to several logarithms of algebraic numbers is in the domain of transcendence theory.

- If the restriction that α and β be algebraic is removed, the statement does not remain true in general. For example,

$$\left(\sqrt{2}^{\sqrt{2}}\right)^{\sqrt{2}} = \sqrt{2}^{\sqrt{2}\cdot\sqrt{2}} = \sqrt{2}^{2} = 2.$$

Here, α is "2^-2 which (as proved by the theorem itself) is transcedental rather than algebraic. Similarly, if you put α = 3 and β = (log 2)/(log 3), which is transcendental, then $\alpha^\beta = 2$ is algebraic. A characterisation of the values for α and β which yield a transcendental α^β is not known.

- Kurt Mahler proved the *p*-adic analogue of the theorem: if α and β are in C_p, the completion of the algebraic closure of Q_p, and they are algebraic over Q, and if |α " $1|_p < 1$ and |β " $1|_p < 1$, then $(\log_p\alpha)/(\log_p\beta)$ is either rational or transcendental, where $\log_p$ is the *p*-adic logarithm function.

Glaisher's Theorem

In number theory, Glaisher's theorem is an identity useful to the study of integer partitions. It is named for James Whitbread Lee Glaisher. It states that the number of partitions of an integer *N* into parts not divisible by *d* is equal to the number of partitions of the form $N = N_1 + \cdots + N_k$ where $N_i \geq N_{i+1}$ and $N_i \geq N_{i+d-1} + 1$, that is, partiti-ons in which no part is repeated *d* or more times.

When $d = 2$ this becomes the special case, known as Euler's theorem, that the number of partitions of into distinct parts is the same as the number of partitions of into odd parts.

Similar Theorems

If instead of counting the number of partitions with distinct parts we count the number of partitions with parts differing by at least 2, a theorem similar to Euler's theorem known as Rogers' theorem (after Leonard James Rogers) is obtained:

The number of partitions whose parts differ by at least 2 is equal to the number of partitions involving only numbers congruent to 1 or 4 (mod 5). For example, there are 6 partitions of 10 into parts differing by at least 2, namely 10, 9+1, 8+2, 7+3, 6+4, 6+3+1; and 6 partitions of 10 involving only 1, 4, 6, 9 ..., namely 9+1, 6+4, 6+1+1+1+1, 4+4+1+1, 4+1+1+1+1+1+1, 1+1+1+1+1+1+1+1+1+1. The theorem was discovered independently by Schur and Ramanujan.

Hasse–Arf Theorem

In mathematics, specifically in local class field theory, the Hasse–Arf theorem is a result concerning jumps of a filtration of the Galois group of a finite Galois extension. A special case of it was originally proved by Helmut Hasse, and the general result was proved by Cahit Arf.

Statement

Higher Ramification Groups

The theorem deals with the upper numbered higher ramification groups of a finite abelian extension L/K. So assume L/K is a finite Galois extension, and that v_K is a discrete normalised valuation of K, whose residue field has characteristic $p > 0$, and which admits a unique extension to L, say w. Denote by v_L the associated normalised valuation ew of L and let be the valuation ring of L under v_L. Let L/K have Galois group G and define the s-th ramification group of L/K for any real $s \geq -1$ by

$$G_s(L/K) = \{\sigma \in G{:}v_L(\sigma a - a) \geq s+1 \text{ for all } a \in \mathcal{O}\}.$$

So, for example, G_{-1} is the Galois group G. To pass to the upper numbering one ha to define the function $\psi_{L/K}$ which in turn is the inverse of the function $\eta_{L/K}$ defined by

$$\eta_{L/K}(s) = \int_0^s \frac{dx}{|G_0 : G_x|}.$$

The upper numbering of the ramification groups is then defined by $G^t(L/K) = G_s(L/K)$ where $s = \psi_{L/K}(t)$.

These higher ramification groups $G^t(L/K)$ are defined for any real $t \geq -1$, but since v_L is a discrete valuation, the groups will change in discrete jumps and not continuously. Thus we say that t is a jump of the filtration $\{G^t(L/K) : t \geq -1\}$ if $G^t(L/K) \neq G^u(L/K)$ for any $u > t$. The Hasse–Arf theorem tells us the arithmetic nature of these jumps

Statement of the Theorem

With the above set up, the theorem states that the jumps of the filtration $\{G^t(L/K) : t \geq -1\}$ are all rational integers.

Hasse–Minkowski Theorem

The Hasse–Minkowski theorem is a fundamental result in number theory which states that two quadratic forms over a number field are equivalent if and only if they are equivalent *locally at all places*, i.e. equivalent over every completion of the field (which may be real, complex, or p-adic). A special case is that a quadratic space over a number field is isotropic if and only if it is isotropic locally everywhere, or equivalently, that a quadratic form over a number field nontrivially represents zero if and only if this holds for all completions of the field. The theorem was proved in the case of the field of rational numbers by Hermann Minkowski and generalized to number fields by Helmut Hasse. The same statement holds even more generally for all global fields.

The importance of the Hasse–Minkowski theorem lies in the novel paradigm it presented for answering arithmetical questions: in order to determine whether an equation of a certain type has a solution in rational numbers, it is sufficient to test whether it has solutions over complete fields of real and p-adic numbers, where analytic considerations, such as Newton's method and its p-adic analogue, Hensel's lemma, apply. This is encapsulated in the idea of a local-global principle, which is one of the most fundamental techniques in arithmetic geometry.

Application to the Classification of Quadratic Forms

The Hasse–Minkowski theorem reduces the problem of classifying quadratic forms over a number field K up to equivalence to the set of analogous but much simpler questions over local fields.

Basic invariants of a nonsingular quadratic form are its dimension, which is a positive integer, and its discriminant modulo the squares in K, which is an element of the multiplicative group K^*/K^{*2}. In addition, for every place v of K, there is an invariant coming from the completion K_v.

- *Case of R.* By Sylvester's law of inertia, the signature (or, alternatively, the negative index of inertia) is a complete invariant.
- *Case of C.* All nonsingular quadratic forms of the same dimension are equivalent.
- *Case of Q_p and its algebraic extensions.* Forms of the same dimension are classified up to equivalence by their Hasse invariant.

These invariants must satisfy some compatibility conditions: a parity relation (the sign of the discriminant must match the negative index of inertia) and a product formula (a local–global relation). Conversely, for every set of invariants satisfying these relations, there is a quadratic form over K with these invariants.

Hilbert's Irreducibility Theorem

In number theory, Hilbert's irreducibility theorem, conceived by David Hilbert, states that every finite number of irreducible polynomials in a finite number of variables and having rational number coefficients admit a common specialisation of a proper subset of the variables to rational numbers such that all the polynomials remain irreducible.

This theorem is a prominent theorem in number theory.

Formulation of the Theorem

Hilbert's irreducibility theorem. Let

$$f_1(X_1,\ldots,X_r,Y_1,\ldots,Y_s),\ldots,f_n(X_1,\ldots,X_r,Y_1,\ldots,Y_s)$$

be irreducible polynomials in the ring

$$\mathbb{Q}[X_1,\ldots,X_r,Y_1,\ldots,Y_s].$$

Then there exists an r-tuple of rational numbers $(a_1,\ldots,a_r)$ such that

$$f_1(a_1,\ldots,a_r,Y_1,\ldots,Y_s),\ldots,f_n(a_1,\ldots,a_r,Y_1,\ldots,Y_s)$$

are irreducible in the ring

$$\mathbb{Q}[Y_1,\ldots,Y_s].$$

Remarks.

- It follows from the theorem that there are infinitely many r-tuples. In fact the set of all irreducible specialisation, called Hilbert set, is large in many senses. For example, this set is Zariski dense in $\mathbb{Q}^r$
- There are always (infinitely many) integer specialisations, i.e., the assertion of the theorem holds even if we demand $(a_1,...,a_r)$ to be integers.
- There are many Hilbertian fields, i.e., fields satisfying Hilbert's irreducibility theorem. For example, global fields are Hilbertian.
- The irreducible specialisation property stated in the theorem is the most general. There are many reductions, e.g., it suffices to take $n = r = s = 1$ in the definition. A recent result of Bary-Soroker shows that for a field K to be Hilbertian it suffices to consider the case of $n = r = s = 1$ and $f = f_1$ absolutely irreducible, that is, irreducible in the ring $K^{alg}[X,Y]$, where K^{alg} is the algebraic closure of K.

Applications

Hilbert's irreducibility theorem has numerous applications in number theory and algebra. For example:

- The inverse Galois problem, Hilbert's original motivation. The theorem almost immediately implies that if a inite group G can be realised as the Galois group of a Galois extension N of

  $$E = \mathbb{Q}(X_1, \ldots, X_r),$$

  then it can be specialised to a Galois extension N_0 of the rational numbers with G as its Galois group.
- Construction of elliptic curves with large rank.
- Hilbert's irreducibility theorem is used as a step in the Andrew Wiles proof of Fermat's last theorem.
- If a polynomial $g(x) \in \mathbb{Z}[x]$ is a perfect square for all large integer values of x, then $g(x)$ is the square of a polynomial in $\mathbb{Z}[x]$. This follows from Hilbert's irreducibility theorem with $n = r = s = 1$ and $f_1(X,Y) = Y^2 - g(X)$.

(More elementary proofs exist.) The same result is true when "square" is replaced by "cube", "fourth power", etc.

Generalisations

It has been reformulated and generalised extensively, by using the language of algebraic geometry.

Hurwitz's Theorem (Number Theory)

In number theory, Hurwitz's theorem, named after Adolf Hurwitz, gives a bound on a Diophantine approximation. The theorem states that for every irrational number ξ there are infinitely many rationals m/n such that

$$\left|\xi - \frac{m}{n}\right| < \frac{1}{\sqrt{5}n^2}.$$

The hypothesis that ξ is irrational cannot be omitted. Moreover the constant is the best possible; if we replace by any number and we let (the golden ratio) then there exist only *finitely* many rational numbers m/n such that the formula above holds.

Jacobi's Four-square Theorem

In 1834, Carl Gustav Jakob Jacobi found an exact formula for the total number of ways a given positive integer n can be represented as the sum of four squares.

Two representations are considered different if their terms are in different order or if the integer being squared (not just the square) is different; to illustrate, these are three of the eight different ways to represent 1:

$$1^2 + 0^2 + 0^2 + 0^2$$
$$0^2 + 1^2 + 0^2 + 0^2$$
$$(-1)^2 + 0^2 + 0^2 + 0^2.$$

The number of ways to represent n as the sum of four squares is eight times the sum of the divisors of n if n is odd and 24 times the sum of the odd divisors of n if n is even, i.e.

$$r_4(n) = \begin{cases} 8\sum_{m|n} m & \text{if } n \text{ is odd} \\ [12pt]24\sum_{\substack{m|n \\ m \text{ odd}}} m & \text{if } n \text{ is even.} \end{cases}$$

Equivalently, it is eight times the sum of all its divisors which are not divisible by 4, i.e.

$$r_4(n) = 8 \sum_{m:4\nmid m|n} m.$$

In particular, for a prime number p we have the explicit formula $r_4(p) = 8(p + 1)$.

The proof shows that the Theta series for the lattice Z^4 is a modular form of a certain level, and hence equals a linear combination of Eisenstein series.

Jurkat–Richert Theorem

The Jurkat–Richert theorem is a mathematical theorem in sieve theory. It is a key ingredient in proofs of Chen's theorem on Goldbach's conjecture. It was proved in 1965 by Wolfgang B. Jurkat and Hans-Egon Richert.

Statement of the Theorem

This formulation is from Diamond & Halberstam. Other formulations are in Jurkat & Richert, Halberstam & Richert, and Nathanson.

Suppose A is a finite sequence of integers and P is a set of primes. Write A_d for the number of items in A that are divisible by d, and write $P(z)$ for the product of the elements in P that are less than z. Write $\omega(d)$ for a multiplicative function such that $\omega(p)/p$ is approximately the proportion of elements of A divisible by p write X for any convenient approximation to $|A|$, and write the remainder as

$$r_A(d) = |A_d| - \frac{\omega(d)}{d} X.$$

Write $S(A,P,z)$ for the number of items in A that are relatively prime to $P(z)$. Write

$$V(z) = \prod_{p \in P, p<z} \left(1 - \frac{\omega(p)}{p}\right).$$

Write $\nu(m)$ for the number of distinct prime divisors of m. Write F_1 and f_1 for functions satisfying certain difference differential equations. We assume the dimension (sifting density) is 1: that is, there is a constant C such that for $2 \le z < w$ we have

$$\prod_{z \le p < w} \left(1 - \frac{\omega(p)}{p}\right)^{-1} \le \left(\frac{\log w}{\log z}\right)\left(1 + \frac{C}{\log z}\right).$$

Then the Jurkat–Richert theorem states that for any numbers y and z with $2 \le z \le y \le X$ we have

$$S(A,P,z) \le XV(z)\left(F_1\left(\frac{\log y}{\log z}\right) + O\left(\frac{(\log\log y)^{3/4}}{(\log y)^{1/4}}\right)\right) + \sum_{m|P(z),m<y} 4^{\nu(m)}|r_A(m)|$$

and

$$S(A,P,z) \ge XV(z)\left(f_1\left(\frac{\log y}{\log z}\right) - O\left(\frac{(\log\log y)^{3/4}}{(\log y)^{1/4}}\right)\right) - \sum_{m|P(z),m<y} 4^{\nu(m)}|r_A(m)|.$$

Kaplansky's Theorem on Quadratic Forms

In mathematics, Kaplansky's theorem on quadratic forms is a result on simultaneous representation of primes by quadratic forms.

It was proved in 2003 by Canadian mathematician Irving Kaplansky (1917–2006).

Statement of the Theorem

Kaplansky's theorem states that a prime p congruent to 1 modulo 16 is representable by both or none of $x^2 + 32y^2$ and $x^2 + 64y^2$, whereas a prime p congruent to 9 modulo 16 is representable by exactly one of these quadratic forms.

This is remarkable since the primes represented by each of these forms individually are *not* describable by congruence conditions.

Proof

Kaplansky's proof uses th facts that 2 is a 4th power modulo p if and only if p is representable by $x^2 + 64y^2$, and that "4 is an 8th power modulo p if and only if p is representable by $x^2 + 32y^2$.

Examples

- The prime $p = 17$ is congruent to 1 modulo 16 and is representable by neither $x^2 + 32y^2$ nor $x^2 + 64y^2$.
- The prime $p=113$ is congruent to 1 modulo 16 and is representable by both $x^2 + 32y^2$ and x^2+64y^2 (since $113 = 9^2 + 32\times1^2$ and $113 = 7^2 + 64\times1^2$).
- The prime $p = 41$ is congruent to 9 modulo 16 and is representable by $x^2 + 32y^2$ (since $41 = 3^2 + 32\times1^2$), but not by $x^2 + 64y^2$.
- The prime $p = 73$ is congruent to 9 modulo 16 and is representable by $x^2 + 64y^2$ (since $73 = 3^2 + 64\times1^2$), but not by $x^2 + 32y^2$.

Similar Results

Five results similar to Kaplansky's theorem are known:

- A prime p congruent to 1 modulo 20 is representable by both or none of $x^2 + 20y^2$ and $x^2 + 100y^2$, whereas a prime p congruent to 9 modulo 20 is representable by exactly one of these quadratic forms.
- A prime p congruent to 1, 16 or 22 modulo 39 is representable by both or none of $x^2 + xy + 10y^2$ and $x^2 + xy + 127y^2$, whereas a prime p congruent to 4, 10 or 25 modulo 39 is representable by exactly one of these quadratic forms.
- A prime p congruent to 1, 16, 26, 31 or 36 modulo 55 is representable by both or none of $x^2 + xy + 14y^2$ and $x^2 + xy\ 69y^2$, whereas a prime p congruent to 4, 9, 14, 34 or 49 modulo 55 is representable by exactly one of these quadratic forms.
- A prime p congruent to 1, 65 or 81 modulo 112 is representable by both or none of $x^2 + 14y^2$ and $x^2 + 448y^2$, whereas a prime p congruent to 9, 25 or 57 modulo 112 is representable by exactly one of these quadratic forms.
- A prime p congruent to 1 or 169 mdulo 240 is representable by both or none of $x^2 + 150y^2$ and $x^2 + 960y^2$, whereas a prime p congruent to 49 or 121 modulo 240 is representable by exactly one of these quadratic forms.

It is conjectured that there are no other similar results involving definite forms.

Kummer's Theorem

In mathematics, Kummer's theorem, named after Ernst Kummer, is as follows. Given integers n e– m e– 0 and a prime number p, the maximum integer k such that p^k divides the binomial coefficient $\binom{n}{m}$ is equal to the number of carries when m is added to $n - m$ in base p.

Lagrange's Four-square Theorem

Lagrange's four-square theorem, also known as Bachet's conjecture, states that any natural number can be represented as the sum of four integer squares

$$p = a_0^2 + a_1^2 + a_2^2 + a_3^2$$

where the four numbers a_0, a_1, a_2, a_3 are integers. For illustration, 3, 31 and 310 can be represented as the sum of four squares as follows:

$$3 = 1^2 + 1^2 + 1^2 + 0^2$$

$$31 = 5^2 + 2^2 + 1^2 + 1^2$$

$$310 = 17^2 + 4^2 + 2^2 + 1^2.$$

This theorem was proven by Joseph Louis Lagrange in 1770, and corresponds to Fermat's theorem on sums of two squares.

Historical development

The theorem appears in the *Arithmetica* of Diophantus, translated into Latin by Bachet in 1621.

Adrien-Marie egendre improved on the theorem in 1798 by stating that a positive integer can be expressed as the sum of three squares if and only if it is not of the form $4^k(8m + 7)$. His proof was incomplete, leaving a gap which was later filled by Carl Friedrich Gauss.

Proof Using the Hurwitz Integers

One of the ways to prove the theorem relies on Hurwitz quaternions, which are the analog of integers for quaternions. The Hurwitz quaternions consist of all quaternions with integer components and all quaternions with half-integer components. These two sets can be combined into a single formula

$$\alpha = \frac{1}{2}E_0(1+\mathbf{i}+\mathbf{j}+\mathbf{k})+E_1\mathbf{i}+E_2\mathbf{j}+E_3\mathbf{k} = a_0 + a_1\mathbf{i}+a_2\mathbf{j}+a_3\mathbf{k}$$

where E_0, E_1, E_2, E_3 are integers. Thus, the quaternion components a_0, a_1, a_2, a_3 are either all integers or all half-integers, depending on whether E_0 is even or odd, respectively. The set of Hurwitz quaternions forms a ring; in particular, the sum or product of any two Hurwitz quaternions is likewise a Hurwitz quaternion.

The generalized uclidean algorithmidentifies the greatest common right or left divisor of two Hurwitz quaternions, where the "size" of the remainder ρ is measured by the norm. The norm N(α) of a quaternion α is the nonnegative real number

$$\mathrm{N}(\alpha) = \alpha\bar{\alpha} = a_0^2 + a_1^2 + a_2^2 + a_3^2$$

where $\bar{\alpha} = a_0 - a_1\mathbf{i} - a_2\mathbf{j} - a_3\mathbf{k}$ is the conjugate of α.

Since quaternion multiplication is associative, and real numbers commute with other quaternions, the norm of a product of quaternions equals the product of the norms:

$$\mathrm{N}(\alpha\beta) = \alpha\beta\overline{(\alpha\beta)} = \alpha\beta\bar{\beta}\bar{\alpha} = \alpha\mathrm{N}(\beta)\bar{\alpha} = \alpha\bar{\alpha}\mathrm{N}(\beta) = \mathrm{N}(\alpha)\mathrm{N}(\beta)$$

The ring of Hurwitz quaternions is Euclidean, since for any quaternion $\alpha = a_0 + a_1\mathbf{i}+a_2\mathbf{j}+a_3\mathbf{k}$ with rational coefficients we can choose a Hurwitz quaternion $\beta = b_0 + b_1\mathbf{i}+b_2\mathbf{j}+b_3\mathbf{k}$ so that $\mathrm{N}(\alpha-\beta) < 1$ by first choosing b_0 so that $|a_0 - b_0| \le 1/4$ and then b_1, b_2, b_3 so that $|a_i - b_i| \le 1/2$ for i1, 2, 3. Then we obtain

$$\mathrm{N}(\alpha-\beta) = (a_0-b_0)^2 + (a_1-b_1)^2 + (a_2-b_2)^2 + (a_3-b_3)^2$$

$$\le \left(\frac{1}{4}\right)^2 + \left(\frac{1}{2}\right)^2 + \left(\frac{1}{2}\right)^2 + \left(\frac{1}{2}\right)^2 = \frac{13}{16} < 1.$$

Therefore, the ring of Hurwitz quaternions is Euclidean and, as a consequence, also a unique factorisation domain.

Since any natural number can be factored into powers of primes, the four-square theorem therefore holds for all natural numbers if it is true for all prime numbers. It is true for 2 =

$0^2 + 0^2 + 1^2 + 1^2$. To show this for an odd prime integer p, represent it as a quaternion (p, 0, 0, 0) and assume that it can be factored into two Hurwitz quaternions

$$p = \alpha\beta.$$

The norms of p, α, β are nonnegative integers such that

$$N(p) = p^2 = N(\alpha\beta) = N(\alpha)N(\beta).$$

But the norm has only two factors, both p. Therefore, if it can be factored into Hurwitz quaternions, then p is the sum of four squares

$$p = N(\alpha) = a_0^2 + a_1^2 + a_2^2 + a_3^2.$$

Lagrange proved that any odd prime p divides at least one number of the form $1 + l^2 + m^2$, where l and m are integers. The latter number can be factored in Hurwitz quaternions:

$$1 + l^2 + m^2 = (1 + l\,\mathrm{i} + m\,\mathrm{j})\,(1 - l\,\mathrm{i} - m\,\mathrm{j}).$$

If p could not be factored in Hurwitz quaternions, it would be a Hurwitz prime number by definition. Then, by unique factorisation, p would have to divide either $1 + l\,\mathrm{i} + m\,\mathrm{j}$ or $1 - l\,\mathrm{i} - m\,\mathrm{j}$ to form another Hurwitz quaternion. But for $p>2$, the number

$$1/p \pm l/p\ \mathrm{i} \pm m/p\ \mathrm{j}$$

is not a Hurwitz integer. Therefore, every $p>2$ can be factored in Hurwitz quaternions, and the four-square theorem holds.

Generalisations

Lagrange's four-square theorem is a special case of the Fermat polygonal number theorem and Waring's problem. Another possible generalisation is the following problem: Given natural numbers a, b, c, d, can we solve

$$n = ax_1^2 + bx_2^2 + cx_3^2 + dx_4^2$$

for all positive integers n in integers x_1, x_2, x_3, x_4? The case $a = b = c = d = 1$ is answered in the positive by Lagrange's four-square theorem. The general solution was given by Ramanujan. He proved that if we assume, without loss of generality, that $a \le b \le c \le d$ then there are exactly 54 possible choices for a, b, c, d such that the problem is solvable in integers x_1, x_2, x_3, x_4 for all n. (Ramanujan listed a 55th possibility $a = 1$, $b = 2$, $c = 5$, $d = 5$, but in this case the problem is not solvable if $n = 15$.)

Algorithms

Michael O. Rabin and Jeffrey Shallit have found randomized polynomial-time algorithms for computing a representation $n = x_1^2 + x_2^2 + x_3^2 + x_4^2$ for a given integer n, in expected running time $O((\log n)^2)$.

Uniqueness

The sequence of positive integers whose representation as a sum of four squares is unique (up to order) is:

1, 2, 3, 5, 6, 7, 8, 11, 14, 15, 23, 24, 32, 56, 96, 128, 224, 384, 512, 896 ... (sequence A006431 in OEIS). These integers consist of the seven odd numbers 1, 3, 5, 7, 11, 15, 23 and all numbers of the form 2×4^k, 6×4^k or 14×4^k. The sequence of positive integers which cannot be represented as a sum of four *non-zero* squares is: 1, 2, 3, 5, 6, 8, 9, 11, 14, 17, 24, 29, 32, 41, 56, 96, 128, 224, 384, 512, 896 ... (sequence A000534 in OEIS). These integers consist of the eight odd numbers 1, 3, 5, 9, 11, 17, 29, 41 and all numbers of the form 2×4^n, 6×4^n or 14×4^n.

Lambek–Moser Theorem

In combinatorial number theory, the Lambek–Moser theorem is a generalisation of Beatty's theorem that defines a partition of the positive integers into two subsets from any monotonic integer-valued function. Conversely, any partition of the positive integers into two subsets may be defined from a monotonic function in this way.

The theorem was discovered by Leo Moser and Joachim Lambek. Dijkstra (1980) provides a visual proof of the result.Statement of the theorem

The theorem applies to any non-decreasing and unbounded function f that maps positive integers to non-negative integers. From any such function f, define f^* to be the integer-valued function that is as close as possible to the inverse function of f, in the sense that, for all n,

$f(f^*(n)) < n \leq f(f^*(n) + 1)$. It follows from this definition that $f^{**} = f$.

Further, define

$F(n) = f(n) + n$ and $G(n) = f^*(n) + n$.

Then the result states that F and G are strictly increasing and that the ranges of F and G form a partition of the positive integers.

Example

Let $f(n) = n^2$;[2] then $f^*(n) = |\sqrt{n-1}|$. Thus $F(n) = n^2 + n$ and $G(n) = |\sqrt{n-1}| + n$. For $n = 1, 2, 3, \ldots$ the values of F are the pronic numbers

2, 6, 12, 20, 30, 42, 56, 72, 90, 110, ...

while the values of G are

1, 3, 4, 5, 7, 8, 9, 10, 11, 13, 14, ...

These two sequences are complementary: each positive integer belongs to exactly one of them.

The Lambek–Moser theorem states that this phenomenon is not specific to the pronic numbers, but rather it arises for any choice of f with the appropriate properties.

Universality

The Lambek–Moser theorem is universal, in the sense that it can explain any partition of the integers into two infinite parts. If $S = s_1, s_2, \ldots$ and $T = t_1, t_2, \ldots$ are any two infinite subsets forming a partition of the integers, one may construct a pair of functions f and f^* from which this partition may be derived using the Lambek–Moser theorem: define $f(n) = s_n - n$ and $f^*(n) = t_n - n$.

For instance, consider the partition of integers into even and odd numbers: let S be the even numbers and T be the odd numbers. Then $s_n = 2n$, so $f(n) = n$ and similarly $f^*(n) = n - 1$. These two functions f and f^* form an inverse pair, and the partition generated via the Lambek–Moser theorem from this pair is just the partition of the positive integers into even and odd numbers.

Lambek and Moser discuss formulas involving the prime-counting function for the functions f and f^* arising in this way from the partition of the positive integers into prime numbers and composite numbers.

Chapter 11

Stochastic Differential Equation

A stochastic differential equation (SDE) is a differential equation in which one or more of the terms is a stochastic process, resulting in a solution which is itself a stochastic process. SDE are used to model diverse phenomena such as fluctuating stock prices or physical system subject to thermal fluctuations. Typically, SDEs incorporate white noise which can be thought of as the derivative of Brownian motion (or the Wiener process); however, it should be mentioned that other types of random fluctuations are possible, such as jump processes.

Background

The earliest work on SDEs was done to describe Brownian motion in Einstein's famous paper, and at the same time by Smoluchowski. However, one of the earlier works related to Brownian motion is credited to Bachelier (1900) in his thesis 'Theory of Speculation'. This work was followed upon by Langevin. Later Itô and Stratonovich put SDEs on more solid mathematical footing.

Terminology

In physical science, SDEs are usually written as Langevin equations. These are sometimes confusingly called "the Langevin equation" even though there are many possible forms. These consist of an ordinary differential equation containing a deterministic part and an additional random white noise term. A second form is the Smoluchowski equation and, more generally, the Fokker-Planck equation. These are partial differential equations that describe the time evolution of probability distribution functions. The third form is the stochastic differential equation that is used most frequently in mathematics and quantitative finance. This is similar to the Langevin form, but it is usually written in differential form. SDEs come in two varieties, corresponding to two versions of stochastic calculus.

Stochastic Calculus

Brownian motion or the Wiener process was discovered to be exceptionally complex mathematically. The Wiener process is nowhere differentiable; thus, it requires its own rules of calculus. There are two dominating versions of stochastic calculus, the Ito stochastic calculus and the Stratonovich stochastic calculus. Each of the two has advantages and disadvantages, and newcomers are often confused whether the one is more appropriate than the other in a given situation. Guidelines exist (e.g. Ø ksendal, 2003) and conveniently, one can readily convert an Ito SDE to an equivalent Stratonovich SDE and back again. Still, one must be careful which calculus to use when the SDE is initially written down.

Numerical Solutions

Numerical solution of stochastic differential equations and especially stochastic partial differential equations is a young field relatively speaking. Almost all algorithms that are used for the solution of ordinary differential equations will work very poorly for SDEs, having very poor numerical convergence.

Methods include the Euler–Maruyama method, Milstein method and Runge–Kutta method (SDE).

Use in Physics

In physics, SDEs are typically written in the Langevin form and referred to as "the Langevin equation." For example, a general coupled set of first-order SDEs is often written in the form:

$$\dot{x}_i = \frac{dx_i}{dt} = f_i(\mathbf{x}) + \sum_{m=1}^{n} g_i^m(\mathbf{x})\eta_m(t),$$

where $\mathbf{x} = \{x_i \mid 1 \le i \le k\}$ is the set of unknowns, the f_i and g_i are arbitrary functions and the η_m are random functions of time, often referred to as "noise terms". This form is usually usable because there are standard techniques for transforming higher-order equations into several coupled first-order equations by introducing new unknowns. If the are constants, the system is said to be subject to additive noise, otherwise it is said to be subject to multiplicative noise. This term is somewhat misleading as it has come to mean the general case even though it appears to imply the limited case where : $g(x) \propto x$. Additive noise is the simpler of the two cases; in that situation the correct solution can often be found using ordinary calculus and in particular the ordinary chain rule of calculus. However, in the case of multiplicative noise, the Langevin equation is not a well-defined entity on its own, and it must be specified whether the Langevin equation should be interpreted as an Ito SDE or a Stratonovich SDE.

In physics, the main method of solution is to find the probability distribution function as a function of time using the equivalent Fokker-Planck equation (FPE). The Fokker-Planck equation is a deterministic partial differential equation. It tells how the probability distribution function evolves in time similarly to how the Schrödinger equation gives the time evolution of the quantum wave function or the diffusion equation gives the time evolution of chemical concentration. Alternatively numerical solutions can be obtained by Monte Carlo simulation. Other techniques include the path integration that draws on the analogy between statistical physics and quantum mechanics (for example, the Fokker-Planck equation can be transformed into the Schrödinger equation by rescaling a few variables) or by writing down ordinary differential equations for the statistical moments of the probability distribution function.

Note on "the Langevin Equation"

The "the" in "the Langevin equation" is somewhat ungrammatical nomenclature. Each individual physical model has its own Langevin equation. Perhaps, "a Langevin equation" or "the associated Langevin equation" would conform better with common English usage.

Use in Probability and Mathematical Finance

The notation used in probability theory (and in many applications of probability theory, for instance mathematical finance) is slightly different. This notation makes the exotic nature of the random function of time η_m in the physics formulation more explicit. It is also the notation used in publications on numerical methods for solving stochastic differential equations. In strict mathematical terms, can not be chosen as a usual function, but only as a generalised function. The mathematical formulation treats this complication with less ambiguity than the physics formulation.

A typical equation is of the form

$$dX_t = \mu(X_t, t)dt + \sigma(X_t, t)dB_t,$$

where denotes a Wiener process (Standard Brownian motion). This equation should be interpreted as an informal way of expressing the corresponding integral equation

$$X_{t+s} - X_t = \int_t^{t+s} \mu(X_u, u)du + \int_t^{t+s} \sigma(X_u, u)dB_u.$$

The equation above characterises the behaviour of the continuous time stochastic process X_t as the sum of an ordinary Lebesgue integral and an Itô integral. A heuristic (but very helpful) interpretation of the stochastic differential equation is that in a small time interval of length δ the stochastic process X_t changes its value by an amount that is normally distributed with expectation $\mu(X_t, t)\ \delta$ and variance $\sigma(X_t, t)^2\ \delta$ and is independent of the past behaviour of the process. This is so because the increments of a Wiener process are independent and normally distributed. The function μ is referred to as the drift coefficient, while σ is called the diffusion coefficient. The stochastic process X_t is called a diffusion process, and is usually a Markov process.

The formal interpretation of an SDE is given in terms of what constitutes a solution to the SDE. There are two main definitions of a solution to an SDE, a strong solution and a weak solution. Both require the existence of a process X_t that solves the integral equation version of the SDE. The difference between the two lies in the underlying probability space (Ω F, Pr). A weak solution consists of a probability space and a process that satisfies the integral equation, while a strong solution is a process that satisfies the equation and is defined on a given probability space. An important example is the equation for geometric Brownian motion

$$dX_t = \mu X_t dt + \sigma X_t dB_t.$$

which is the equation for the dynamics of the price of a stock in the Black Scholes options pricing model of financial mathematics.There are also more general stochastic differential equations where the coefficients μ and σ depend not only on the present value of the process X_t, but also on previous values of the process and possibly on present or previous values of other processes too.

In that case the solution process, *X*, is not a Markov process, and it is called an Itô process and not a diffusion process. When the coefficients depends only on present and past values of *X*, the defining equation is called a stochastic delay differential equation.

Delay Differential Equation

In mathematics, delay differential equations (DDEs) are a type of differential equation in which the derivative of the unknown function at a certain time is given in terms of the values of the function at previous times.

A general form of the time-delay differential equation for $x(t) \in \mathbb{R}^n$ is

$$\frac{d}{dt}x(t) = f(t, x(t), x_t),$$

where $x_t = \{x(\tau) : \tau \le t\}$ represents the trajectory of the solution in the past. In this equation, *f* is a functional operator from $\mathbb{R} \times \mathbb{R}^n \times C^1$ to $\mathbb{R}^n$.

Examples

Continuous delay

$$\frac{d}{dt}x(t) = f\left(t, x(t), \int_{-\infty}^{0} x(t+\tau)d\mu(\tau)\right)$$

Discrete delay

$$\frac{d}{dt}x(t) = f(t, x(t), x(t-\tau_1), \ldots, x(t-\tau_m))$$

for $\tau_1 > \cdots > \tau_m \ge 0$.

Linear with discrete delay

$$\frac{\mathrm{d}}{\mathrm{d}t}x(t) = A_0 x(t) + A_1 x(t-\tau_1) + \cdots + A_m x(t-\tau_m)$$

where $A_0, \ldots, A_m \in \mathbb{R}^{n\times n}$.

Pantograph equation

$$\frac{\mathrm{d}}{\mathrm{d}t}x(t) = ax(t) + bx(\lambda t),$$

where a, b and λ are constants and $0 < \lambda < 1$. This equation and some more general forms are named after the pantographs on trains.

Solving DDEs

DDEs are mostly solved in a stepwise fashion with a principle called the method of steps. For instance, consider the DDE with a single delay

$$\frac{\mathrm{d}}{\mathrm{d}t}x(t) = f(x(t), x(t-\tau))$$

with given initial condition $\phi : [-\tau, 0] \to \mathbb{R}^n$. Then the solution on the interval $[0, \tau]$ is given by $\psi(t)$ which is the solution to the inhomogeneous initial value problem

$$\frac{\mathrm{d}}{\mathrm{d}t}\psi(t) = f(\psi(t), \phi(t-\tau))$$

with $\psi(0) = \phi(0)$. This can be continued for the successive intervals by using the solution to the previous interval as inhomogeneous term. In practice, the initial value problem is often solved numerically.

Reduction to ODE

In some cases, delay differential equations are equivalent to a system of ordinary differential equations.

- Example 1

  Consider an equation $\frac{\mathrm{d}}{\mathrm{d}t}x(t) = f\left(t, x(t), \int_{-\infty}^{0} x(t+\tau)e^{\lambda\tau}\,\mathrm{d}\tau\right)$.

  Introduce $y(t) = \int_{-\infty}^{0} x(t+\tau)e^{\lambda\tau}\,\mathrm{d}\tau$ to get a system of ODEs

  $$\frac{\mathrm{d}}{\mathrm{d}t}x(t) = f(t, x, y), \quad \frac{\mathrm{d}}{\mathrm{d}t}y(t) = x - \lambda y.$$

- Example 2

  An equation $\frac{d}{dt}x(t) = f\left(t, x(t), \int_{-\infty}^{0} x(t+\tau)\cos(\alpha\tau+\beta)d\tau\right)$ is equivalent to

$$\frac{d}{dt}x(t) = f(t,x,y), \quad \frac{d}{dt}y(t) = \cos(\beta)x + \alpha z, \quad \frac{d}{dt}z(t) = \sin(\beta)x - \alpha y,$$

where

$$y = \int_{-\infty}^{0} x(t+\tau)\cos(\alpha\tau+\beta)d\tau, \quad z = \int_{-\infty}^{0} x(t+\tau)\sin(\alpha\tau+\beta)d\tau.$$

The Characteristic Equation

Similar to ODEs, many properties of linear DDEs can be characterised and analysed using the *characteristic equation.* The characteristic equation associated with the linear DDE with discrete delays

$$\frac{d}{dt}x(t) = A_0 x(t) + A_1 x(t-\tau_1) + \cdots + A_m x(t-\tau_m)$$

is

$$\det(-\lambda I + A_0 + A_1 e^{-\tau_1\lambda} + \cdots + A_m e^{-\tau_m\lambda}) = 0$$

The roots λ of the characteristic equation are called characteristic roots or eigenvalues and the solution set is often referred to as the spectrum. Because of the exponential in the characteristic equation, the DDE has, unlike the ODE case, an infinite number of eigenvalues, making a spectral analysis more involved.

The spectrum does however have a some properties which can be exploited in the analysis. For instance, even though there are an infinite number of eigenvalues, there are only a finite number of eigenvalues to the right of any vertical line in the complex plane.

This characteristic equation is a nonlinear eigenproblem and there are many methods to compute the spectrum numerically. In some special situations it is possible to solve the characteristic equation explicitly. Consider, for example, the following DDE:

$$\frac{d}{dt}x(t) = -x(t-1).$$

The characteristic equation is

$$-\lambda - e^{-\lambda} = 0.$$

There are an infinite number of solutions to this equation for complex λ. They are given by

$$\lambda = W_k(-1)$$

where W_k is the kth branch of the Lambert W function.

Solutions

In mathematics, in the study of differential equations, the Picard–Lindelöf theorem, Picard's existence theorem or Cauchy–Lipschitz theorem is an important theorem on existence and uniqueness of solutions to first-order equations with given initial conditions.

The theorem is named after Charles Émile Picard, Ernst Lindelöf, Rudolf Lipschitz and Augustin-Louis Cauchy.

Picard–Lindelöf Theorem

Consider the initial value problem

$$y'(t) = f(t, y(t)), \quad y(t_0) = y_0, \quad t \in [t_0 - \varepsilon, t_0 + \varepsilon].$$

Suppose f is Lipschitz continuous in y and continuous in t. Then, for some value $\epsilon > 0$, there exists a unique solution $y(t)$ to the initial value problem within the range $[t_0 - \epsilon, t_0 + \epsilon]$.

Proof Sketch

The proof relies on transforming the differential equation, and applying fixed-point theory. By integrating both sides, any function satisfying the differential equation must also satisfy the integral equation

$$y(t) - y(t_0) = \int_{t_0}^{t} f(s, y(s))ds.$$

A simple proof of existence of the solution is obtained by successive approximations. In this context, the method is known as Picard iteration.

Set

$$\varphi_0(t) = y_0$$

and

$$\varphi_{k+1}(t) = y_0 + \int_{t_0}^{t} f(s, \varphi_k(s))ds.$$

It can then be shown, by using the Banach fixed point theorem, that the sequence of "Picard iterates" φ_k is convergent and that the limit is a solution to the problem.

Exploiting the fact that the width of the interval where the local solution is defined is entirely determined by the Lipschitz constant of the function, one can assure global existence of the solution, i.e. the solution exists and is unique until it leaves the domain of definition of the ODE.

An application of Grönwall's lemma to $|\varphi(t) - \psi(t)|$, where φ and ψ are two solutions, shows that $\varphi(t) = \psi(t)$, thus proving the global uniqueness (the local uniqueness is a consequence of the uniqueness of the Banach fixed point).

Detailed Proof

Let $C_{a,b} = \overline{I_a(t_0)} \times \overline{B_b(y_0)}$ be the compact cylinder where f is defined, this is

$$t \in \overline{I_a(t_0)} = [t_0 - a, t_0 + a]$$

and

$$\overline{B_b(y_0)} = [y_0 - b, y_0 + b].$$

Let

$$M = \sup_{C_{a,b}} ||f||,$$

this is, the maximum slope of the function in modulus. Finally, let L be the Lipschitz constant of with respect to the second variable. We will proceed to apply Banach fixed point theorem using the metric on $\mathcal{C}(I_a(t_0), B_b(y_0))$ induced by the uniform norm

$$\|\varphi_1\|_\infty = \sup_{t \in I_a} |\varphi(t)|.$$

We define an operator between two functional spaces of continuous functions, Picard's operator, as follows:

$$\Gamma : \mathcal{C}(I_a(t_0), B_b(y_0)) \rightarrow \mathcal{C}(I_a(t_0), B_b(y_0))$$

defined by:

$$\Gamma\varphi(t) = y_0 + \int_{t_0}^{t} f(s, \varphi(s))ds.$$

We impose that it is well-defined, in other words, that its image must be a function taking values on $B_b(y_0)$, or equivalently, that the norm of

$\Gamma\varphi(t) - y_0$ is less than b,

which can be restated as

$$\|\varphi_1\|_\infty \leq b.$$

$$|\,|\Gamma\varphi(t) - y_0\,|\,| = |\,|\int_{t_0}^{t} f(s, \varphi(s))ds\,|\,| \leq |\int_{t_0}^{t} |\,|$$

$$f(s, \varphi(s))\,|\,|ds\,| \leq M\,|\,t - t_0\,| \leq Ma \leq b$$

The last step is the imposition, so we are impose the requirement

$$a < b / M.$$

Let us impose now the Picard's operator to be contractive under certain hypothesis over α that later on we will be able to omit.

Given two functions $\varphi_1, \varphi_2 \in \mathcal{C}(I_\alpha(t_0), B_b(y_0))$, in order to apply the Banach fixed point theorem we want

$$\|\Gamma\varphi_1 - \Gamma\varphi_2\|_\infty \leq q\|\varphi_1 - \varphi_2\|_\infty, \text{ for some } q < 1.$$

So let t be such that

$$\|\Gamma\varphi_1 - \Gamma\varphi_2\|_\infty = \|(\Gamma\varphi_1 - \Gamma\varphi_2)(t)\|$$

then using the definition of Γ

$$\|\ \|(\Gamma\varphi_1 - \Gamma\varphi_2)(t)\| =$$

$$\left\|\int_{t_0}^{t} (f(s, \varphi_1(s)) - f(s, \varphi_2(s))ds)\right\| \leq \left|\int_{t_0}^{t} |\,|\,f(s, \varphi_1(s)) - f(s, \varphi_2(s))\,|\,|ds\right|.$$

$$\varphi_1 - \Gamma\varphi_2\,\|_\infty = \|(\Gamma\varphi_1 - \Gamma\varphi_2)(t)\|$$

Then since f is Lipschitz with respect to the second variable, we have that:

$$L\,|\int_{t_0}^{t} |\,|\,\varphi_1(s) - \varphi_2(s)\,|\,|ds\,| \leq La\,|\,|\,\varphi_1 - \varphi_2\,|\,|.$$

This is contractive if $a < 1/L$. We have established that the Picard's operator is a contraction on the Banach spaces with the metric induced by the uniform norm. This allows us to apply the Banach fixed point theorem to conclude that the operator has a unique fixed point. In particular, there is a unique function

$$\varphi \in \mathcal{C}(I_a(t_0), B_b(y_0)) \text{ such that } \Gamma\varphi = \varphi.$$

This function is the unique solution of the initial value problem, valid on the interval I_a where a satisfies the condition

$$a < \min\{b/M, 1/L\}.$$

Other Existence Theorems

The Picard–Lindelöf theorem shows that the solution exists and that it is unique. The Peano existence theorem shows only existence, not uniqueness, but it assumes only that f is continuous in y, instead of Lipschitz continuous. For example, the right-hand side of the equation $y\,2 = y^{1/3}$ with initial condition $y(0) = 0$ is continuous but not Lipschitz continuous. Indeed, the solution of this equation is not unique; two different solutions are given besides the trivial one $y(t) = 0$

$$y(t) = \pm\left(\tfrac{2}{3}t\right)^{3/2}.$$

Even more general is Carathéodory's existence theorem, which proves existence (in a more general sense) under weaker conditions on f.

Wronskian

In mathematics, the Wronskian is a determinant introduced by Jσzef Hoene-Wronski (1812) and named by Thomas Muir (1882Chapter XVIII). It is used in the study of, differential equations, where it can sometimes be used to show that a set of solutions is linearly independent.

Definition

The Wronskian of two functions f and g is $W(f,g) = fg2 - gf\,2$.

More generally, for n real- or complex-valued functions $f_1, \ldots, f_n$, which are $n - 1$ times differentiable on an interval I, the Wronskian $W(f_1, \ldots, f_n)$ as a function on I is defined by

$$W(f_1,\ldots,f_n)(x)=\begin{vmatrix} f_1(x) & f_2(x) & \cdots & f_n(x) \\ f_{1'}(x) & f_{2'}(x) & \cdots & f_{n'}(x) \\ \vdots & \vdots & \ddots & \vdots \\ f_1^{(n-1)}(x) & f_2^{(n-1)}(x) & \cdots & f_n^{(n-1)}(x) \end{vmatrix}, \qquad x\in I.$$

That is, it is the determinant of the matrix constructed by placing the functions in the first row, the first derivative of each function in the second row, and so on through the (n - 1)st derivative, thus forming a square matrix sometimes called a fundamental matrix.

When the functions f_i are solutions of a linear differential equation, the Wronskian can be found explicitly using Abel's identity, even if the functions f_i are not known explicitly.

The Wronskian and Linear Independence

If the functions f_i are linearly dependent, then so are the columns of the Wronskian as differentiation is a linear operation, so the Wronskian vanishes. So the Wronskian can be used to show that a set of differentiable functions is linearly independent on an interval by showing that it does not vanish identically.

A common misconception is that $W = 0$ everywhere implies linear dependence, but Peano (1889) pointed out that the functions x^2 and $|x|x$ have continuous derivatives and their Wronskian vanishes everywhere, yet they are not linearly dependent in any neighbourhood of 0. There are several extra conditions which ensure that the vanishing of the Wronskian in an interval implies linear dependence.

Peano (1889) observed that if the functions are analytic, then the vanishing of the Wronskian in an interval implies that they are linearly dependent. Bochner (1901) gave several other conditions for the vanishing of the Wronskian to imply linear dependence; for example, if the Wronskian of n functions is identically zero and the n Wronskians of n–1 of them do not all vanish at any point then the functions are linearly dependent. Wolsson (1989a) gave a more general condition that together with the vanishing of the Wronskian implies linear dependence.

Generalised Wronskians

For n functions of several variables, a generalised Wronskian is the determinant of an n by n matrix with entries

$D_i(f_j)$ (with 0d–i<n), where each D_i is some constant coefficient linear partial differential operator of order i. If the functions are linearly dependent then all generalised Wronskians vanish. As in the 1 variable case the converse is not true in general: if all generalised Wronskians vanish this does not imply that the functions are linearly dependent. However the converse is true in many special cases. For example, if the functions are polynomials and all generalised Wronskians vanish, then the functions are linearly dependent. Roth used this result about generalised Wronskians in his proof of Roth's theorem.

Phase Portrait

A phase portrait is a geometric representation of the trajectories of a dynamical system in the phase plane. Each set of initial conditions is represented by a different curve, or point. Phase portraits are an invaluable tool in studying dynamical systems. They consist of a plot of typical trajectories in the state space. This reveals information such as whether an attractor, a repellor or limit cycle is present for the chosen parameter value. The concept of topological equivalence is important in classifying the behaviour of systems by specifying when two different phase portraits represent the same qualitative dynamic behaviour.

A phase portrait graph of a dynamical system depicts the system's trajectories (with arrows) and stable steady states (with dots) and unstable steady states (with circles) in a state space. The axes are of state variables.

Phase Space

In mathematics and physics, a phase space, introduced by Willard Gibbs in 1901, is a space in which all possible states of a system are represented, with each possible state of the system corresponding to one unique point in the phase space. For mechanical systems, the phase space usually consists of all possible values of position and momentum variables i.e. the cotangent space of configuration space.

A plot of position and momentum variables as a function of time is sometimes called a phase plot or a phase diagram. Phase diagram, however, is more usually reserved in the physical sciences for a diagram showing the various regions of stability of the thermodynamic phases of a chemical system, which consists of pressure, temperature, and composition.

In a phase space, every degree of freedom or parameter of the system is represented as an axis of a multidimensional space; a one-dimensional system is called a phase line, while a two-dimensional system is called a phase plane. For every possible state of the system, or allowed combination of values of the system's parameters, a point is plotted in the multidimensional space. Often this succession of plotted points is analogous to the system's state evolving over time. In the end, the phase diagram represents all that the system can be, and its shape can easily elucidate qualities of the system that might not be obvious otherwise. A phase space may contain very many dimensions. For instance, a gas containing many molecules may require a separate dimension for each particle's x, y and z positions and momenta as well as any number of other properties.

In classical mechanics, any choice of generalised coordinates q^i for the position (i.e. coordinates on configuration space) defines conjugate generalised momenta p_i which together define co-ordinates on phase space. More abstractly, in classical mechanics phase space is the cotangent space of configuration space, and in this interpretaton the procedure above expresses that a choice of local coordinates on configuration space induces a choice of natural local Darboux coordinates for the standard symplectic structure on a cotangent space.

The motion of an ensemble of systems in this space is studied by classical statistical mechanics. The local density of points in such systems obeys Liouville's Theorem, and so can be taken as constant. Within the context of a model system in classical mechanics, the phase space coordinates of the system at any given time are composed of all of the system's dynamical variables. Because of this, it is possible to calculate the state of the system at any given time in the future or the past, through integration of Hamilton's or Lagrange's equations of motion.

Examples

Low Dimensions: For simple systems, there may be as few as one or two degrees of freedom. One degree of freedom occurs when one has an autonomous ordinary differential equation in a single variable, $dy/dt = f(y)$, with the resulting one-dimensional system being called a phase line, and the qualitative behaviour of the system being immediately visible from the phase line. The simplest non-trivial examples are the exponential growth model/decay (one unstable/stable equilibrium) and the logistic growth model (two equilibria, one stable, one unstable).

The phase space of a two-dimensional system is called a phase plane, which occurs in classical mechanics for a single particle moving in one dimension, and where the two variables are position and velocity. In this case, a sketch of the phase portrait may give qualitative information about the dynamics of the system, such as the limit-cycle of the Van der Pol oscillator shown in the diagram.

Here, the horizontal axis gives the position and vertical axis the velocity. As the system evolves, its state follows one of the lines (trajectories) on the phase diagram.

Chaos Theory

Classic examples of phase diagrams from chaos theory are :

- the Lorenz attractor
- population growth (i.e. Logistic map)
- parameter plane of complex quadratic polynomials with Mandelbrot set.

Quantum Mechanics

In quantum mechanics, the coordinates p and q of phase space normally become hermitian operators in a Hilbert space.

But they may alternatively retain their classical interpretation, provided functions of them compose in novel algebraic ways (through Groenewold's 1946 star product), consistent with the uncertainty principle of quantum mechanics. Every quantum mechanical observable corresponds to a unique function or distribution on phase space, and vice versa, as specified by Hermann Weyl (1927) and supplemented by John von Neumann (1931); Eugene Wigner (1932); and, in a grand synthesis, by H J Groenewold (1946). With J E Moyal (1949), these completed the foundations of the phase space formulation of quantum mechanics, a complete and logically autonomous reformulation of quantum mechanics. (Its modern abstractions include deformation quantisation and geometric quantisation.)

Expectation values in phase-space quantisation are obtained isomorphically to tracing operator observables with the density matrix in Hilbert space: they are obtained by phase-space integrals of observables, with the Wigner quasi-probability distribution effectively serving as a measure.

Thus, by expressing quantum mechanics in phase space (the same ambit as for classical mechanics), the Weyl map facilitates recognition of quantum mechanics as a deformation (generalisation) of classical mechanics, with deformation parameter '/*S*, where *S* is the action of the relevant process. (Other familiar deformations in physics involve the deformation of classical Newtonian into relativistic mechanics, with deformation parameter v/c; or the deformation of Newtonian gravity into General Relativity, with deformation parameter Schwarzschild-radius/characteristic-dimension.)

Classical expressions, observables, and operations (such as Poisson brackets) are modified by '-dependent quantum corrections, as the conventional commutative multiplication applying in classical mechanics is generalised to the noncommutative star-multiplication characterising quantum mechanics and underlying its uncertainty principle.

Thermodynamics and Statistical Mechanics

In thermodynamics and statistical mechanics contexts, the term phase space has two meanings: It is used in the same sense as in classical mechanics. If a thermodynamical system consists of *N* particles, then a point in the *6N*-dimensional phase space describes the dynamical state of every particle in that system, as each particle is associated with three position variables and three momentum variables. In this sense, a point in phase space is said to be a microstate of the system. *N* is typically on the order of Avogadro's number, thus describing the system at a microscopic level is often impractical. This leads us to the use of phase space in a different sense.

The phase space can refer to the space that is parametrised by the *macroscopic* states of the system, such as pressure, temperature, etc. For instance, one can view the pressure-volume diagram or entropy-temperature diagrams as describing part of this phase space. A point in this phase space is correspondingly called a macrostate. There may easily be more than one microstate with the same macrostate. For example, for a fixed temperature, the system could have many dynamic configurations at the microscopic level. When used in this sense, a phase is a region of phase space where the system in question is in, for example, the liquid phase, or solid phase, etc.

Since there are many more microstates than macrostates, the phase space in the first sense is usually a manifold of much larger dimensions than the second sense. Clearly, many more parameters are required to register every detail of the system down to the molecular or atomic scale than to simply specify, say, the temperature or the pressure of the system.

Phase Integral

In classical statistical mechanics (continuous energies) the concept of phase space provides a classical analog to the partition function (sum over states) known as the phase integral. Instead of summing the Boltzmann factor over discretely spaced energy states (defined by appropriate integer quantum numbers for each degree of freedom) one may integrate over continuous phase space. Such integration essentially consists of two parts: integration of the momentum component of all degrees of freedom (momentum space) and integration of the position component of all degrees of freedom (configuration space). Once the phase integral is known, it may be related to the classical partition function by multiplication of a normalisation constant representing the number of quantum energy states per unit phase space. As shown in detail in, this normalisation constant is simply the inverse of Planck's constant raised to a power equal to the number of degrees of freedom for the system.

Lyapunov Stability

Various types of stability may be discussed for the solutions of differential equations describing dynamical systems. The most important type is that concerning the stability of solutions near to a point of equilibrium. This may be discussed by the theory of Lyapunov. In simple terms, if all solutions of the dynamical system that start out near an equilibrium point x_e stay near forever, then is Lyapunov stable. More strongly, if is Lyapunov stable and all solutions that start out near converge to, then is asymptotically stable. The notion of exponential stability guarantees a minimal rate of decay, i.e., an estimate of how quickly the solutions converge. The idea of Lyapunov stability can be extended to infinite-dimensional manifolds, where it is known as structural stability, which concerns the behaviour of different but "nearby" solutions to differential equations. Input-to-state stability (ISS) applies Lyapunov notions to systems with inputs.

History

Lyapunov stability is named after Aleksandr Lyapunov, a Russian mathematician who published his book "The General Problem of Stability of Motion" in 1892. Lyapunov was the first to consider the modifications necessary in nonlinear systems to the linear theory of stability based on linearising near a point of equilibrium. His work, initially published in Russian and then translated to French, received little attention for many years. Interest in it started suddenly during the Cold War (1953-1962) period when the so-called "Second Method of Lyapunov" was found to be applicable to the stability of aerospace guidance systems which typically contain strong nonlinearities not treatable by other methods.

A large number of publications appeared then and since in the control and systems literature. More recently the concept of the Lyapunov exponent (related to Lyapunov's First Method of discussing stability) has received wide interest in connection with chaos theory. Lyapunov stability methods have also been applied to finding equilibrium solutions in traffic assignment problems.

Lyapunov's Second Method for Stability

Lyapunov, in his original 1892 work, proposed two methods for demonstrating stability. The first method developed the solution in a series which was then proved convergent within limits. The second method, which is almost universally used nowadays, makes use of a *Lyapunov function V(x)* which has an analogy to the potential function of classical dynamics. It is introduced as follows for a system having a point of equilibrium at x = 0. Consider a function $V(x):\mathbb{R}^n \to \mathbb{R}$ such that

- $V(x) \geq 0$ with equality if and only if $x = 0$ (positive definite)
- $\dot{V}(x) = \frac{d}{dt}V(x) \leq 0$ with equality if and only if $x = 0$ (negative definite).

Then *V(x)* is called a Lyapunov function candidate and the system is asymptotically stable in the sense of Lyapunov (i.s.L.).

(Note that $V(0) = 0$ is required; otherwise for example $V(x) = 1/(1+|x|)$ would "prove" that $\dot{x}(t) = x$ is locally stable. An additional condition called "properness" or "radial unboundedness" is required in order to conclude global asymptotic stability.)

It is easier to visualise this method of analysis by thinking of a physical system (e.g. vibrating spring and mass) and considering the energy of such a system. If the system loses energy over time and the energy is never restored then eventually the system must grind to a stop and reach some final resting state. This final state is called the attractor.

However, finding a function that gives the precise energy of a physical system can be difficult, and for abstract mathematical systems, economic systems or biological systems, the concept of energy may not be applicable. Lyapunov's realisation was that stability can be proven without requiring knowledge of the true physical energy, provided a Lyapunov function can be found to satisfy the above constraints.

Definition for Discrete-time Systems

The definition for discrete-time systems is almost identical to that for continuous-time systems. The definition below provides this, using an alternate language commonly used in more mathematical texts.

Let (X,d) be a metric space and $f : X \to X$ a continuous function. A point $x \in X$ is said to be Lyapunov stable, if, for each $\epsilon > 0$, there is a $\delta > 0$ such that for all $y \in X$, if

$$d(x, y) < \delta$$

then

$$d(f^n(x), f^n(y)) < \epsilon$$

for all $n \in \mathbb{N}$.

We say that x is asymptotically stable if it belongs to the interior of its stable set, *i.e.* if there is a such that

$$\lim_{n\to\infty} d(f^n(x), f^n(y)) = 0$$

whenever $d(x,y) < \delta$.

Stability for Linear State Space Models

A linear state space model

$$\dot{\mathbf{x}} = A\mathbf{x}$$

is asymptotically stable (in fact, exponentially stable) if all real parts of the eigenvalues of A are negative. This condition is equivalent to the following one:

$$A^T M + MA + N = 0$$

has a solution where $N = N^T > 0$ and $M = M^T > 0$ (positive definite matrices). (The relevant Lyapunov function is $V(x) = x^T Mx$.)

Correspondingly, a time-discrete linear state space model

$$\mathbf{x}_{t+1} = A\mathbf{x}_t$$

is asymptotically stable (in fact, exponentially stable) if all the eigenvalues of A have a modulus smaller than one.

This latter condition has been generalised to switched systems: a linear switched discrete time system (ruled by a set of matrices $\{A_1,\ldots,A_m\}$)

$$\mathbf{x}_{t+1} = A_{i_t}\mathbf{x}_t, \quad A_{i_t} \in \{A_1,\ldots,A_m\}$$

is asymptotically stable (in fact, exponentially stable) if the joint spectral radius of the set is smaller than one.

Stability for Systems with Inputs

A system with inputs (or controls) has the form

$$\dot{\mathbf{x}} = f(\mathbf{x},\mathbf{u})$$

where the (generally time-dependent) input u(t) may be viewed as a *control*, *external input*, *stimulus*, *disturbance*, or *forcing function*. The study of such systems is the subject of control theory and applied in control engineering. For systems with inputs, one must quantify the effect of inputs on the stability of the system. The main two approaches to this analysis are

BIBO stability (for linear systems) and input-to-state (ISS) stability (for nonlinear systems)

Example

Consider an equation, where compared to the Van der Pol oscillator equation the friction term is changed:

$$\ddot{y} + y - \varepsilon\left(\frac{\dot{y}^3}{3} - \dot{y}\right) = 0.$$

The equilibrium is at : $\ddot{y} = y = 0$. Here is a good example of an unsuccessful try to find a Lyapunov function that proves stability:

Let

$$x_1 = y, x_2 = \dot{y}$$

so that the corresponding system is

$$\dot{x}_2 = -x_1 + \varepsilon\left(\frac{{x_2}^3}{3} - x_2\right).$$

Let us choose as a Lyapunov function

$$V = \frac{1}{2}\left(x_1^2 + x_2^2\right)$$

which is clearly positive definite. Its derivative is

$$\begin{aligned} \dot{V} &= x_1\dot{x}_1 + x_2\dot{x}_2 \\ &= x_1x_2 - x_1x_2 + \varepsilon\left(\frac{x_2^4}{3} - x_2^2\right) \\ &= -\varepsilon\left(x_2^2 - \frac{x_2^4}{3}\right). \end{aligned}$$

It seems that if the parameter ε is positive, stability is asymptotic for $x_2^2 < 3$. But this is wrong, since $\dot{V}$ does not depend on x_1, and will be 0 everywhere on the axis.

Exponential Stability

In control theory, a continuous linear time-invariant system is exponentially stable if and only if the system has eigenvalues (i.e., the poles of input-to-output systems) with strictly negative real parts. (i.e., in the left half of the complex plane). A discrete-time input-to-output LTI system is exponentially stable if and only if the poles of its transfer function lie strictly within the unit circle centred on the origin of the complex plane. Exponential stability is a form of asymptotic stability. Systems that are not LTI are exponentially stable if their convergence is bounded by exponential decay.

Practical Consequences

An exponentially stable LTI system is one that will not "blow up" (i.e., give an unbounded output) when given a finite input or non-zero initial condition. Moreover, if the system is given a fixed, finite input (i.e., a step), then any resulting oscillations in the output will decay at an exponential rate, and the output will tend asymptotically to a new final, steady-state value. If the system is instead given a Dirac delta impulse as input, then induced oscillations will die away and the system will return to its previous value. If oscillations do not die away, or the system does not return to its original output when an impulse is applied, the system is instead marginally stable.

Example Exponentially Stable LTI Systems

The graph on the right shows the impulse response of two similar systems. The green curve is the response of the system with impulse response $y(t) = e^{-\frac{t}{5}}$, while the blue represents the system $y(t) = e^{-\frac{t}{5}} \sin(t)$. Although one response is oscillatory, both return to the original value of 0 over time.

Real-world Example

Imagine putting a marble in a ladle. It will settle itself into the lowest point of the ladle and, unless disturbed, will stay there. Now imagine giving the ball a push, which is an approximation to a Dirac delta impulse.

The marble will roll back and forth but eventually resettle in the bottom of the ladle. Drawing the horizontal position of the marble over time would give a gradually diminishing sinusoid rather like the blue curve in the image above.

A step input in this case requires supporting the marble away from the bottom of the ladle, so that it cannot roll back.

It will stay in the same position and will not, as would be the case if the system were only marginally stable or entirely unstable, continue to move away from the bottom of the ladle under this constant force equal to its weight.

It is important to note that in this example the system is not stable for all inputs. Give the marble a big enough push, and it will fall out of the ladle and fall, stopping only when it reaches the floor. For some systems, therefore, it is proper to state that a system is exponentially stable *over a certain range of inputs*.

Rate of Convergence

In numerical analysis, the speed at which a convergent sequence approaches its limit is called the rate of convergence. Although strictly speaking, a limit does not give information about any finite first part of the sequence, this concept is of practical importance if we deal with a sequence of successive approximations for an iterative method, as then typically fewer iterations are needed to yield a useful approximation if the rate of convergence is higher. This may even make the difference between needing ten or a million iterations.

Similar concepts are used for discretisation methods. The solution of the discretised problem converges to the solution of the continuous problem as the grid size goes to zero, and the speed of convergence is one of the factors of the efficiency of the method. However, the terminology in this case is different from the terminology for iterative methods.

Series acceleration is a collection of techniques for improving the rate of convergence of a series discretisation. Such acceleration is commonly accomplished with sequence transformations.

Convergence Speed for Iterative Methods

Basic Definition: Suppose that the sequence $\{x_k\}$ converges to the number L. We say that this sequence converges linearly to L, if there exists a number μ – (0, 1) such that

$$\lim_{k\to\infty} \frac{|x_{k+1} - L|}{|x_k - L|} = \mu.$$

The number μ is called the *rate of convergence.*

If the sequences converges, and

- $\mu = 0$, then the sequence is said to converge superlinearly.
- $\mu = 1$, then the sequence is said to converge sublinearly.

If the sequences *converges sublinearly* and additionally

$$\lim_{k\to\infty} \frac{|x_{k+2} - x_{k+1}|}{|x_{k+1} - x_k|} = 1,$$

then it is said the sequence $\{x_k\}$ converges logarithmically to L.

The next definition is used to distinguish superlinear rates of convergence. We say that the sequence converges with order q to L for $q>1$ if

$$\lim_{k\to\infty} \frac{|x_{k+1} - L|}{|x_k - L|^q} = \mu| \; \mu > 0.$$

In particular, convergence with order

- 2 is called quadratic convergence,
- 3 is called cubic convergence,
- etc.

This is sometimes called *Q-linear convergence*, *Q-quadratic convergence*, etc., to distinguish it from the definition below. The Q stands for "quotient," because the definition uses the quotient between two successive terms.

Extended Definition: The drawback of the above definitions is that these do not catch some sequences which still converge reasonably fast, but whose "speed" is variable, such as the sequence $\{b_k\}$ below.

Therefore, the definition of rate of convergence is sometimes extended as follows.

Under the new definition, the sequence $\{x_k\}$ converges with at least order q if there exists a sequence $\{\varepsilon_k\}$ such that

$$|x_k - L| \le \varepsilon_k^q \quad \text{for all } k,$$

and the sequence $\{\varepsilon_k\}$ converges to zero with order q according to the above "simple" definition.

To distinguish it from that definition, this is sometimes called *R-linear convergence*, *R-quadratic convergence*, etc. (with the R standing for "root").

Convergence Speed for Discretisation Methods

A similar situation exists for discretisation methods. The important parameter here for the convergence speed is not the iteration number k but it depends on the number of grid points and grid spacing. In this case, the number of grid points in a discretisation process is inversely proportional to the grid spacing here it denoted as n.

In this case, a sequence x_n is said to converge to L with order p if there exists a constant C such that

$$|x_n - L| < Cn^{-p} \text{ for all } n.$$

This is written as $|x_n - L| = O(n^{-p})$ using the big O notation. This is the relevant definition when discussing methods for numerical quadrature or the solution of ordinary differential equations.

Examples

The sequence $\{d_k\}$ with $d_k = 1 / (k+1)$ was introduced above. This sequence converges with order 1 according to the convention for discretisation methods.

The sequence $\{a_k\}$ with $a_k = 2^{-k}$, which was also introduced above, converges with order p for every number p. It is said to converge exponentially using the convention for discretisation methods. However, it only converges linearly (that is, with order 1) using the convention for iterative methods.

Acceleration of Convergence

Many methods exist to increase the rate of convergence of a given sequence, i.e. to transform a given sequence into one converging faster to the same limit. Such techniques are in general known as "series acceleration". The goal of the transformed sequence is to be much less "expensive" to calculate than the original sequence. One example of series acceleration is Aitken's delta-squared process.

Power Series Solution of Differential Equations

In mathematics, the power series method is used to seek a power series solution to certain differential equations. In general, such a solution assumes a power series with unknown coefficients, then substitutes that solution into the differential equation to find a recurrence relation for the coefficients.

Method

Consider the second-order linear differential equation

$$a_2(z)f''(z)+a_1(z)f'(z)+a_0(z)f(z)=0.$$

Suppose a_2 is nonzero for all z. Then we can divide throughout to obtain

$$f''+\frac{a_1(z)}{a_2(z)}f'+\frac{a_0(z)}{a_2(z)}f=0.$$

Suppose further that a_1/a_2 and a_0/a_2 are analytic functions.

The power series method calls for the construction of a power series solution

$$f=\sum_{k=0}^{\infty}A_k z^k.$$

If a_2 is zero for some z, then the Frobenius method, a variation on this method, is suited to deal with so called *singular points*. The method works analogously for higher order equations as well as for systems.

Chapter 12

Eigenvalues and Eigenvectors of Matrices

Characteristic Polynomial

The eigenvalues of A are precisely the solutions λ to the equation $\det(A-\lambda I)=0$. Here det is the determinant of the matrix formed by A - λI and I is the $n \times n$ identity matrix. This equation is called the *characteristic equation* (or, less often, the secular equation) of A. For example, if A is the following matrix (a so-called diagonal matrix):

$$A = \begin{bmatrix} a_{1,1} & 0 & \cdots & 0 \\ 0 & a_{2,2} & \cdots & 0 \\ \vdots & \vdots & \ddots & 0 \\ 0 & 0 & 0 & a_{n,n} \end{bmatrix},$$

then the characteristic equation reads

$$\det(A-\lambda I) = \det\left(\begin{bmatrix} a_{1,1} & 0 & \cdots & 0 \\ 0 & a_{2,2} & \cdots & 0 \\ \vdots & \vdots & \ddots & 0 \\ 0 & 0 & 0 & a_{n,n} \end{bmatrix} - \lambda \begin{bmatrix} 1 & 0 & \cdots & 0 \\ 0 & 1 & \cdots & 0 \\ \vdots & \vdots & \ddots & 0 \\ 0 & 0 & 0 & 1 \end{bmatrix}\right)$$

$$= \det \begin{bmatrix} a_{1,1}-\lambda & 0 & \cdots & 0 \\ 0 & a_{2,2}-\lambda & \cdots & 0 \\ \vdots & \vdots & \ddots & 0 \\ 0 & 0 & 0 & a_{n,n}-\lambda \end{bmatrix}$$

$$= (a_{1,1}-\lambda)(a_{2,2}-\lambda)\cdots(a_{n,n}-\lambda) = 0$$

The solutions to this equation are the eigenvalues $\lambda_i = a_{i,i}$ (i = 1, ..., n).

Proving the aforementioned relation of eigenvalues and solutions of the characteristic equation requires some linear algebra, specifically the notion of linearly independent vectors: briefly, the eigenvalue equation for a matrix A can be expressed as

$$Ax - \lambda Ix = 0,$$

which can be rearranged to

$$(A - \lambda I)x = 0.$$

If there exists an inverse

$$(A - \lambda I)^{-1},$$

then both sides can be left-multiplied by it, to obtain x = 0. Therefore, if λ is such that $A - \lambda I$ is invertible, λ cannot be an eigenvalue. It can be shown that the converse holds, too: if $A - \lambda I$ is not invertible, λ is an eigenvalue. A criterion from linear algebra states that a matrix (here: $A - \lambda I$) is non-invertible if and only if its determinant is zero, thus leading to the characteristic equation.

The left-hand side of this equation can be seen (using Leibniz' rule for the determinant) to be a polynomial function in λ, whose coefficients depend on the entries of A. This polynomial is called the *characteristic polynomial*. Its degree is n, that is to say, the highest power of λ occurring in this polynomial is λ^n. At least for small matrices, the solutions of the characteristic equation (hence, the eigenvalues of A) can be found directly. Moreover, it is important for theoretical purposes, such as the Cayley–Hamilton theorem. It also shows that any $n \times n$ matrix has at most n eigenvalues. However, the characteristic equation need not have n distinct solutions. In other words, there may be strictly less than n distinct eigenvalues. This happens for the matrix describing the shear mapping discussed below.

If the matrix has real entries, the coefficients of the characteristic polynomial are all real. However, the roots are not necessarily real; they may include complex numbers with a non-zero imaginary component. For example, a 2 × 2 matrix describing a 45° rotation will not leave any non-zero vector pointing in the same direction. However, there is at least one *complex number* λ solving the characteristic equation, even if the entries of the matrix A are complex numbers to begin with. (This existence of such a solution is known as the fundamental theorem of algebra.) For a complex eigenvalue, the corresponding eigenvectors also have complex components.

Eigenspace

If x is an eigenvector of the matrix A with eigenvalue λ, then any scalar multiple αx is also an eigenvector of A with the same eigenvalue, since $A(\alpha x) = \alpha Ax = \alpha\lambda x = \lambda(\alpha x)$. More generally, any non-zero linear combination of eigenvectors that share the same eigenvalue λ, will itself be an eigenvector with eigenvalue λ. Together with the zero vector, the eigenvectors of A with the same eigenvalue form a linear subspace called an *eigenspace* E_λ of the vector space. In case of $\dim(E_\lambda) = 1$, it is called an *eigenline* and λ is called a *scaling factor*.

Diagonalizable matrices can be decomposed into a direct sum of eigenspaces, as per the eigendecomposition of a matrix. If a matrix is not diagonalizable, then it is called defective, and, while it cannot be decomposed into eigenspaces, it can be decomposed into the more general concept of generalised eigenspaces, as discussed here.

Algebraic and Geometric Multiplicities

Given an $n \times n$ matrix A and an eigenvalue λ_i of this matrix, there are two numbers measuring, roughly speaking, the number of eigenvectors belonging to λ_i. They are called *multiplicities*: the *algebraic multiplicity* of an eigenvalue is defined as the multiplicity of the corresponding root of the characteristic polynomial. The *geometric multiplicity* of an eigenvalue is defined as the dimension of the associated eigenspace, i.e. number of linearly independent eigenvectors with that eigenvalue. Both algebraic and geometric multiplicity are integers between (including) 1 and n. The algebraic multiplicity n_i and geometric multiplicity m_i may or may not be equal, but we always have m_i dŠ n_i. The simplest case is of course when $m_i = n_i = 1$. The total number of linearly independent eigenvectors, N_x, is given by summing the geometric multiplicities

$$\sum_{i=1}^{N_\lambda} m_i = N_{\mathbf{x}}.$$

Over a complex vector space, the sum of the algebraic multiplicities will equal the dimension of the vector space, but the sum of the geometric multiplicities may be smaller. In this case, it is possible that there may not be sufficient eigenvectors to span the entire space – more formally, there is no basis of eigenvectors (an eigenbasis). A matrix is diagonalisable by a suitable choice of coordinates if and only if there is an eigenbasis; if a matrix is not diagonalisable, it is said to be defective. For defective matrices, the notion of eigenvector can be generalised to generalised eigenvectors, and over an algebraically closed field a basis of *generalised* eigenvectors always exists, as follows from Jordan form.

The eigenvectors corresponding to different eigenvalues are linearly independent, meaning, in particular, that in an n-dimensional space the linear transformation A cannot have more than n eigenvalues (or eigenspaces).

All defective matrices have fewer than n distinct eigenvalues, but not all matrices with fewer than n distinct eigenvalues are defective – for example, the identity matrix is diagonalisable (and indeed diagonal in any basis), but only has the eigenvalue 1.

Given an ordered choice of linearly independent eigenvectors, especially an eigenbasis, they can be indexed by eigenvalues, *i.e.* using a double index, with $x_{i,j}$ being the jth eigenvector for the ith eigenvalue. The eigenvectors can also be indexed using the simpler notation of a single index x_k, with $k = 1, 2, ..., N_x$.

Worked Example

These concepts are explained for the matrix

$$A = \begin{bmatrix} 2 & 1 \\ 1 & 2 \end{bmatrix}.$$

The characteristic equation of this matrix reads

$$\det(A - \lambda I) = \det \begin{bmatrix} 2-\lambda & 1 \\ 1 & 2-\lambda \end{bmatrix} = 0.$$

Calculating the determinant, this yields the quadratic equation

$$\lambda^2 - 4\lambda + 3 = 0,$$

whose solutions (also called roots) are $\lambda = 1$ and $\lambda = 3$. The eigenvectors for the eigenvalue are determined by using the eigenvalue equation, which in this case reads

$$\begin{bmatrix} 2 & 1 \\ 1 & 2 \end{bmatrix} \begin{bmatrix} x \\ y \end{bmatrix} = 3 \begin{bmatrix} x \\ y \end{bmatrix}.$$

The juxtaposition at the left hand side denotes matrix multiplication. Spelling this out, this equation comparing two vectors is tantamount to a system of the following two linear equations:

$$2x + y = 3x$$
$$x + 2y = 3y.$$

Both equations reduce to the single linear equation $x = y$. That is to say, any vector of the form (x, y) with $y = x$ is an eigenvector to the eigenvalue $\lambda = 3$.

However, the vector (0, 0) is excluded. A similar calculation shows that the eigenvectors corresponding to the eigenvalue $\lambda = 1$ are given by non-zero vectors (x, y) such that $y = -x$.

For example, an eigenvector corresponding to is whereas an eigenvector corresponding to is . These vectors, placed as columns in a matrix, may be used to create a diagonalisable matrix.

Eigendecomposition

Let A be a square $n \times n$ matrix. Let $q_1 \ldots q_k$ be an eigenvector basis, i.e. an indexed set of k linearly independent eigenvectors, where k is the dimension of the space spanned by the eigenvectors of A. If $k = n$, then A can be written

$$A = Q \Lambda Q^{-1}$$

where Q is the square $n \times n$ matrix whose i-th column is the basis eigenvector q_i of A and $\wedge$ is the diagonal matrix whose diagonal elements are the corresponding eigenvalues, i.e. $\wedge_{ii} = \lambda_i$.

Further Properties

Let A be an $n \times n$ matrix with eigenvalues $\lambda_i, i = 1, 2, \ldots, n$.

Then

- Trace of A $\text{tr}(A) = \sum \lambda_i = \lambda_1 + \lambda_2 + \cdots + \lambda_n$
- Determinant of A $\det(A) = \prod \lambda_i = \lambda_1 \lambda_2 \cdots \lambda_n$
- Eigenvalues of A^k are $\lambda_1^k, \ldots, \lambda_n^k$

These first three results follow by putting the matrix in upper-triangular form, in which case the eigenvalues are on the diagonal and the trace and determinant are respectively the sum and product of the diagonal.

- If $A = A^H$, i.e., is Hermitian, every eigenvalue is real.
- Every eigenvalue of a Unitary matrix has absolute value .

Calculation

The complexity of the problem for finding roots/eigenvalues of the characteristic polynomial increases rapidly with increasing the degree of the polynomial (the dimension of the vector space). There are exact solutions for dimensions below 5, but for dimensions greater than or equal to 5 there are generally no exact solutions and one has to resort to numerical methods to find them approximately.

(In fact, since the roots of *any* polynomial can be expressed as eigenvalues of a companion matrix, the Abel–Ruffini theorem implies that there is no general algebraic solution for eigenvalues of 5×5 or larger matrices: any general eigenvalue algorithm is necessarily approximate, although in practice one can obtain any desired accuracy.)

Worse, any computational procedure that starts by computing the coefficients of the characteristic polynomial can be very inaccurate in the presence of round-off error, because the roots of a polynomial are an extremely sensitive function of the coefficients.

Efficient, accurate methods to compute eigenvalues and eigenvectors of arbitrary matrices were not known until the advent of the QR algorithm in 1961. Combining Householder transformation with LU decomposition offers better convergence than the QR algorithm. For large Hermitian sparse matrices, the Lanczos algorithm is one example of an efficient iterative method to compute eigenvalues and eigenvectors, among several other possibilities.

History

Eigenvalues are often introduced in the context of linear algebra or matrix theory. Historically, however, they arose in the study of quadratic forms and differential equations. Euler studied the rotational motion of a rigid body and discovered the importance of the principal axes. Lagrange realised that the principal axes are the eigenvectors of the inertia matrix. In the early 19th century, Cauchy saw how their work could be used to classify the quadric surfaces, and generalised it to arbitrary dimensions. Cauchy also coined the term *racine caractéristique* (characteristic root) for what is now called *eigenvalue*; his term survives in *characteristic equation.*

Fourier used the work of Laplace and Lagrange to solve the heat equation by separation of variables in his famous 1822 book *Théorie analytique de la chaleur*. Sturm developed Fourier's ideas further and brought them to the attention of Cauchy, who combined them with his own ideas and arrived at the fact that real symmetric matrices have real eigenvalues. This was extended by Hermite in 1855 to what are now called Hermitian matrices. Around the same time, Brioschi proved that the eigenvalues of orthogonal matrices lie on the unit circle, and Clebsch found the corresponding result for skew-symmetric matrices.

Finally, Weierstrass clarified an important aspect in the stability theory started by Laplace by realising that defective matrices can cause instability. In the meantime, Liouville studied eigenvalue problems similar to those of Sturm; the discipline that grew out of their work is now called *Sturm–Liouville theory.* Schwarz studied the first eigenvalue of Laplace's equation on general domains towards the end of the 19th century, while Poincaré studied Poisson's equation a few years later.

At the start of the 20th century, Hilbert studied the eigenvalues of integral operators by viewing the operators as infinite matrices. He was the first to use the German word *eigen* to denote eigenvalues and eigenvectors in 1904, though he may have been following a related usage by Helmholtz. For some time, the standard term in English was "proper value", but the more distinctive term "eigenvalue" is standard today.

The first numerical algorithm for computing eigenvalues and eigenvectors appeared in 1929, when Von Mises published the power method. One of the most popular methods today, the QR algorithm, was proposed independently by John G.F. Francis and Vera Kublanovskaya in 1961.

Generalisations

Left and Right Eigenvectors

The word eigenvector formally refers to the right eigenvector x_R. It is defined by the eigenvalue equation

$$Ax_R = \lambda x_R,$$

and is the most commonly used eigenvector. However, the left eigenvector x_L exists as well, and is defined by

$$x_L A = \lambda x_L.$$

The left and the right eigenvalues are the same. However, the left and right eigenvectors are usually different. But in the case of Hermitian (or real symmetric) matrix A, the left and right eigenvectors are equal.

Infinite-Dimensional Spaces and Spectral Theory

If the vector space is an infinite dimensional Banach space, the notion of eigenvalues can be generalised to the concept of spectrum. The spectrum is the set of scalars λ for which $(T - \lambda I)^{-1}$ is not defined; that is, such that $T - \lambda I$ has no bounded inverse.

Clearly if λ is an eigenvalue of T, λ is in the spectrum of T. In general, the converse is not true. There are operators on Hilbert or Banach spaces that have no eigenvectors at all. This can be seen in the following example. The bilateral shift on the Hilbert space $\ell^2(\mathbf{Z})$ (that is, the space of all sequences of scalars ... a_{-1}, a_0, a_1, a_2, ... such that $\cdots + |a_{-1}|^2 + |a_0|^2 + |a_1|^2 + |a_2|^2 + \cdots$ converges) has no eigenvalue but does have spectral values.

In infinite-dimensional spaces, the spectrum of a bounded operator is always nonempty. This is also true for an unbounded self adjoint operator. Via its spectral measures, the spectrum of any self adjoint operator, bounded or otherwise, can be decomposed into absolutely continuous, pure point, and singular parts.

The hydrogen atom is an example where both types of spectra appear. The eigenfunctions of the hydrogen atom Hamiltonian are called eigenstates and are grouped into two categories. The bound states of the hydrogen atom correspond to the discrete part of the spectrum (they have a discrete set of eigenvalues that can be computed by Rydberg formula) while the ionization processes are described by the continuous part (the energy of the collision/ionization is not quantised).

Eigenfunctions

A common example of such maps on infinite dimensional spaces are the action of differential operators on function spaces. As an example, on the space of infinitely differentiable functions, the process of differentiation defines a linear operator since

$$\frac{d}{dt}(af+bg)=a\frac{df}{dt}+b\frac{dg}{dt},$$

where $f(t)$ and $g(t)$ are differentiable functions, and a and b are constants.

The eigenvalue equation for linear differential operators is then a set of one or more differential equations. The eigenvectors are commonly called eigenfunctions. The simplest case is the eigenvalue equation for differentiation of a real valued function by a single real variable. We seek a function (equivalent to an infinite-dimensional vector) that, when differentiated, yields a constant times the original function. In this case, the eigenvalue equation becomes the linear differential equation

$$\frac{d}{dt}f(x)=\lambda f(x).$$

Here λ is the eigenvalue associated with the function, $f(x)$. This eigenvalue equation has a solution for any value of λ. If λ is zero, the solution is

$$f(x)=A,$$

where A is any constant; if λ is non-zero, the solution is the exponential function

$$f(x)=Ae^{\lambda x}.$$

If we expand our horizons to complex valued functions, the value of λ can be any complex number. The spectrum of d/dt

is therefore the whole complex plane. This is an example of a continuous spectrum.

The shape of a standing wave in a string fixed at its boundaries is an example of an eigenfunction of a differential operator.

The admittable eigenvalues are governed by the length of the string and determine the frequency of oscillation.

The displacement, $h(x,t)$, of a stressed rope fixed at both ends, like the vibrating strings of a string instrument, satisfies the wave equation

$$\frac{\partial^2 h}{\partial t^2} = c^2 \frac{\partial^2 h}{\partial x^2},$$

which is a linear partial differential equation, where c is the constant wave speed. The normal method of solving such an equation is separation of variables. If we assume that h can be written as the product of the form $X(x)T(t)$, we can form a pair of ordinary differential equations:

$$X'' = -\frac{\omega^2}{c^2} X \text{and} T'' = -\omega^2 T.$$

Each of these is an eigenvalue equation (the unfamiliar form of the eigenvalue is chosen merely for convenience). For any values of the eigenvalues, the eigenfunctions are given by

$$X = \sin\left(\frac{\omega x}{c} + \phi\right) \text{and} T = \sin(\omega t + \psi).$$

If we impose boundary conditions (that the ends of the string are fixed with $X(x) = 0$ at $x = 0$ and $x = L$, for example) we can constrain the eigenvalues. For those boundary conditions, we find

$$\sin(\phi) = 0 \text{ , and so the phase angle } \phi = 0 \text{ and } \sin\left(\frac{\omega L}{c}\right) = 0.$$

Thus, the constant ω is constrained to take one of the values $\omega_n = \frac{nc\pi}{L}$, where n is any integer. Thus the clamped string supports a family of standing waves of the form

$$h(x,t) = \sin(n\pi x / L)\sin(\omega_n t).$$

From the point of view of our musical instrument, the frequency ω_n is the frequency of the *n*th harmonic, which is called the *(n-1)*st overtone.

Associative Algebras and Representation Theory

More algebraically, rather than generalising the vector space to an infinite dimensional space, one can generalise the algebraic object that is acting on the space, replacing a single operator acting on a vector space with an algebra representation – an associative algebra acting on a module.

The study of such actions is the field of representation theory. To understand these representations, one breaks them into indecomposable representations, and, if possible, into irreducible representations; these correspond respectively to generalised eigenspaces and eigenspaces, or rather the indecomposable and irreducible components of these. While a single operator on a vector space can be understood in terms of eigenvectors – 1-dimensional invariant subspaces – in general in representation theory the building blocks (the irreducible representations) are higher-dimensional.

A closer analog of eigenvalues is given by the notion of a *weight,* with the analogs of eigenvectors and eigenspaces being *weight vectors* and *weight spaces.* For an associative algebra *A* over a field F, the analog of an eigenvalue is a one-dimensional representation $\lambda : A \to \mathbf{F}$ (a map of algebras; a linear functional that is also multiplicative), called the *weight,* rather than a single scalar. A map of algebras is used because if a vector is an eigenvector for two elements of an algebra, then it is also an eigenvector for any linear combination of these, and the eigenvalue is the corresponding linear combination of the eigenvalues, and likewise for multiplication. This is related to the classical eigenvalue as follows: a single operator *T* corresponds to the algebra F[*T*] (the polynomials in *T*), and a map of algebras $\mathbf{F}[T] \to \mathbf{F}$ is determined by its value on the generator *T;* this value is the eigenvalue. A vector *v* on which the algebra acts by this weight (i.e., by scalar multiplication, with the scalar determined by the weight) is called a *weight vector,* and other concepts generalise similarly. The generalisation of a diagonalisable matrix (having an eigenbasis) is a *weight module.*

Because a weight is a map to a field, which is commutative, the map factors through the abelianisation of the algebra A – equivalently, it vanishes on the derived algebra – in terms of matrices, if v is a common eigenvector of operators T and U, then $TUv = UTv$ (because in both cases it is just multiplication by scalars), so common eigenvectors of an algebra must be in the set on which the algebra acts commutatively (which is annihilated by the derived algebra).

Thus of central interest are the free commutative algebras, namely the polynomial algebras. In this particularly simple and important case of the polynomial algebra $\mathbf{F}[T_1,...,T_k]$ in a set of commuting matrices, a weight vector of this algebra is a simultaneous eigenvector of the matrices, while a weight of this algebra is simply a k-tuple of scalars $\lambda = (\lambda_1,...,\lambda_k)$ corresponding to the eigenvalue of each matrix, and hence geometrically to a point in k-space.

These weights – in particularly their geometry – are of central importance in understanding the representation theory of Lie algebras, specifically the finite-dimensional representations of semisimple Lie algebras.

As an application of this geometry, given an algebra that is a quotient of a polynomial algebra on k generators, it corresponds geometrically to an algebraic variety in k-dimensional space, and the weight must fall on the variety – i.e., it satisfies defining equations for the variety. This generalises the fact that eigenvalues satisfy the characteristic polynomial of a matrix in one variable.

Applications

Schrödinger Equation]

The wavefunctions associated with the bound states of an electron in a hydrogen atom can be seen as the eigenvectors of the hydrogen atom Hamiltonian as well as of the angular momentum operator. They are associated with eigenvalues interpreted as their energies (increasing downward: $n = 1,2,3,...$) and angular momentum (increasing across: s, p, d,...). The illustration shows the square of the absolute value of the wavefunctions. Brighter areas correspond to higher probability density for a position measurement. The centre of each figure is the atomic nucleus, a proton.

An example of an eigenvalue equation where the transformation T is represented in terms of a differential operator is the time-independent Schrödinger equation in quantum mechanics:

$$H\psi_E = E\psi_E$$

where H, the Hamiltonian, is a second-order differential operator and ψ_E, the wavefunction, is one of its eigenfunctions corresponding to the eigenvalue E, interpreted as its energy.

However, in the case where one is interested only in the bound state solutions of the Schrödinger equation, one looks for within the space of square integrable functions. Since this space is a Hilbert space with a well-defined scalar product, one can introduce a basis set in which and H can be represented as a one-dimensional array and a matrix respectively. This allows one to represent the Schrödinger equation in a matrix form.

Bra-ket notation is often used in this context. A vector, which represents a state of the system, in the Hilbert space of square integrable functions is represented by . In this notation, the Schrödinger equation is:

$$H|\Psi_E\rangle = E|\Psi_E\rangle$$

where $|\Psi_E\rangle$ is an eigenstate of H. It is a self adjoint operator, the infinite dimensional analog of Hermitian matrices. As in the matrix case, in the equation above $H|\Psi_E\rangle$ is understood to be the vector obtained by application of the transformation H to $|\Psi_E\rangle$.

Molecular Orbitals

In quantum mechanics, and in particular in atomic and molecular physics, within the Hartree–Fock theory, the atomic and molecular orbitals can be defined by the eigenvectors of the Fock operator. The corresponding eigenvalues are interpreted as ionization potentials via Koopmans' theorem. In this case, the term eigenvector is used in a somewhat more general meaning, since the Fock operator is explicitly dependent on the orbitals and their eigenvalues. If one wants to underline this aspect one speaks of nonlinear eigenvalue problem. Such equations are usually solved by an iteration procedure, called in this case self-consistent field method. In quantum chemistry, one often represents the Hartree–Fock equation in a non-orthogonal basis set. This particular representation is a generalised eigenvalue problem called Roothaan equations.

Geology and Glaciology

In geology, especially in the study of glacial till, eigenvectors and eigenvalues are used as a method by which a mass of information of a clast fabric's constituents' orientation and dip can be summarised in a 3-D space by six numbers.

In the field, a geologist may collect such data for hundreds or thousands of clasts in a soil sample, which can only be compared graphically such as in a Tri-Plot (Sneed and Folk) diagram, or as a Stereonet on a Wulff Net. The output for the orientation tensor is in the three orthogonal (perpendicular) axes of space.

Eigenvectors output from programs such as Stereo32 are in the order $E_1 \geq E_2 \geq E_3$, with E_1 being the primary orientation of clast orientation/dip, E_2 being the secondary and E_3 being the tertiary, in terms of strength. The clast orientation is defined as the eigenvector, on a compass rose of 360°. Dip is measured as the eigenvalue, the modulus of the tensor: this is valued from 0° (no dip) to 90° (vertical). The relative values of E_1, E_2, and E_3 are dictated by the nature of the sediment's fabric. If $E_1 = E_2 = E_3$, the fabric is said to be isotropic. If $E_1 = E_2 > E_3$ the fabric is planar. If $E_1 > E_2 > E_3$ the fabric is linear.

Principal Components Analysis

The eigendecomposition of a symmetric positive semidefinite (PSD) matrix yields an orthogonal basis of eigenvectors, each of which has a nonnegative eigenvalue. The orthogonal decomposition of a PSD matrix is used in multivariate analysis, where the sample covariance matrices are PSD. This orthogonal decomposition is called principal components analysis (PCA) in statistics. PCA studies linear relations among variables. PCA is performed on the covariance matrix or the correlation matrix (in which each variable is scaled to have its sample variance equal to one). For the covariance or correlation matrix, the eigenvectors correspond to principal components and the eigenvalues to the variance explained by the principal components. Principal component analysis of the correlation matrix provides an orthonormal eigen-basis for the space of the observed data: In this basis, the largest eigenvalues correspond to the principal-components that are associated with most of the covariability among a number of observed data.

Principal component analysis is used to study large data sets, such as those encountered in data mining, chemical research, psychology, and in marketing. PCA is popular especially in psychology, in the field of psychometrics.

In Q-methodology, the eigenvalues of the correlation matrix determine the Q-methodologist's judgment of *practical* significance (which differs from the statistical significance of hypothesis testing): The factors with eigenvalues greater than 1.00 are considered practically significant, that is, as explaining an important amount of the variability in the data, while eigenvalues less than 1.00 are considered practically insignificant, as explaining only a negligible portion of the data variability. More generally, principal component analysis can be used as a method of factor analysis in structural equation modeling.

Vibration Analysis

Eigenvalue problems occur naturally in the vibration analysis of mechanical structures with many degrees of freedom. The eigenvalues are used to determine the natural frequencies (or eigenfrequencies) of vibration, and the eigenvectors determine the shapes of these vibrational modes. In particular, undamped vibration is governed by

$$m\ddot{x} + kx = 0$$

or

$$m\ddot{x} = -kx$$

that is, acceleration is proportional to position (i.e., we expect x to be sinusoidal in time). In n dimensions, m becomes a mass matrix and k a stiffness matrix. Admissible solutions are then a linear combination of solutions to the generalised eigenvalue problem

$$-kx = \omega^2 mx$$

where ω^2 is the eigenvalue and ω is the angular frequency. Note that the principal vibration modes are different from the principal compliance modes, which are the eigenvectors of k alone. Furthermore, damped vibration, governed by

$$m\ddot{x} + c\dot{x} + kx = 0$$

leads to what is called a so-called quadratic eigenvalue problem,

$$(\omega^2 m + \omega c + k)x = 0$$

This can be reduced to a generalised eigenvalue problem by clever algebra at the cost of solving a larger system.

The orthogonality properties of the eigenvectors allows decoupling of the differential equations so that the system can be represented as linear summation of the eigenvectors. The eigenvalue problem of complex structures is often solved using finite element analysis, but neatly generalise the solution to scalar-valued vibration problems.

Eigenfaces

In image processing, processed images of faces can be seen as vectors whose components are the brightnesses of each pixel. The dimension of this vector space is the number of pixels.

The eigenvectors of the covariance matrix associated with a large set of normalised pictures of faces are called eigenfaces; this is an example of principal components analysis. They are very useful for expressing any face image as a linear combination of some of them.

In the facial recognition branch of biometrics, eigenfaces provide a means of applying data compression to faces for identification purposes. Research related to eigen vision systems determining hand gestures has also been made.

Similar to this concept, eigenvoices represent the general direction of variability in human pronunciations of a particular utterance, such as a word in a language. Based on a linear combination of such eigenvoices, a new voice pronunciation of the word can be constructed. These concepts have been found useful in automatic speech recognition systems, for speaker adaptation.

Tensor of Inertia

In mechanics, the eigenvectors of the inertia tensor define the principal axes of a rigid body. The tensor of inertia is a key quantity required to determine the rotation of a rigid body around its centre of mass.

Stress Tensor

In solid mechanics, the stress tensor is symmetric and so can be decomposed into a diagonal tensor with the eigenvalues on the diagonal and eigenvectors as a basis. Because it is diagonal, in this orientation, the stress tensor has no shear components; the components it does have are the principal components.

Eigenvalues of a Graph

In spectral graph theory, an eigenvalue of a graph is defined as an eigenvalue of the graph's adjacency matrix A, or (increasingly) of the graph's Laplacian matrix, which is either $T-A$ (sometimes called the Combinatorial Laplacian) or $I-T^{-1/2}AT^{-1/2}$ (sometimes called the Normalised Laplacian), where T is a diagonal matrix with $T_{v,v}$ equal to the degree of vertex v, and in $T^{-1/2}$, the vth diagonal entry is $\deg(v)^{-1/2}$. The kth principal eigenvector of a graph is defined as either the eigenvector corresponding to the kth largest or k^{th} smallest eigenvalue of the Laplacian. The first principal eigenvector of the graph is also referred to merely as the principal eigenvector.

The principal eigenvector is used to measure the centrality of its vertices. An example is Google's PageRank algorithm. The principal eigenvector of a modified adjacency matrix of the World Wide Web graph gives the page ranks as its components. This vector corresponds to the stationary distribution of the Markov chain represented by the row-normalised adjacency matrix; however, the adjacency matrix must first be modified to ensure a stationary distribution exists. The second smallest eigenvector can be used to partition the graph into clusters, via spectral clustering. Other methods are also available for clustering.

Cramer's Rule

For a matrix equation

$$Ax = b$$

the solution is given by Cramer's rule:

$$x_i = \frac{\det(A_i)}{\det(A)} \qquad i = 1, \ldots, n$$

where A_i is the matrix formed by replacing the i-th column of A by the column vector b. This follows immediately by column expansion of the determinant, i.e.

$$\det(A_i) = \det\left[a_1, \ \ldots, \ b, \ \ldots, \ a_n\right]$$

$$= \sum_{j=1}^{n} x_j \det\left[a_1, \ \ldots, a_{i-1}, \ a_j, \ a_{i+1}, \ \ldots, \ a_n\right]$$

$$= x_i \det(A)$$

where the vectors a_j are the columns of A. The rule is also implied by the identity

$$A\operatorname{adj}(A) = \operatorname{adj}(A)A = \det(A)I_n.$$

It has recently been shown that Cramer's rule can be implemented in $O(n^3)$ time, which is comparable to more common methods of solving systems of linear equations, such as LU, QR, or singular value decomposition.

Block Matrices

Suppose A, B, C, and D are $n\times n$-, $n\times m$-, $m\times n$-, and $m\times m$-matrices, respectively. Then

$$\det\begin{pmatrix} A & 0 \\ C & D \end{pmatrix} = \det\begin{pmatrix} A & B \\ 0 & D \end{pmatrix} = \det(A)\det(D).$$

This can be seen from the Leibniz formula or by induction on n. When A is invertible, employing the following identity

$$\begin{pmatrix} A & B \\ C & D \end{pmatrix} = \begin{pmatrix} A & 0 \\ C & I \end{pmatrix}\begin{pmatrix} I & A^{-1}B \\ 0 & D-CA^{-1}B \end{pmatrix}$$

leads to

$$\det\begin{pmatrix} A & B \\ C & D \end{pmatrix} = \det(A)\det(D-CA^{-1}B).$$

When D is invertible, a similar identity with $\det(D)$ factored out can be derived analogously, that is,

$$\det\begin{pmatrix} A & B \\ C & D \end{pmatrix} = \det(D)\det(A-BD^{-1}C).$$

When the blocks are square matrices of the same order further formulas hold. For example, if C and D commute (i.e., $CD = DC$), then the following formula comparable to the determinant of a 2-by-2 matrix holds:

$$\det\begin{pmatrix} A & B \\ C & D \end{pmatrix} = \det(AD - BC).$$

Derivative

By definition, e.g., using the Leibniz formula, the determinant of real (or analogously for complex) square matrices is a polynomial function from $\mathrm{R}^{n\times n}$ to R. As such it is everywhere differentiable. Its derivative can be expressed using Jacobi's formula:

$$\frac{\mathrm{d}\det(A)}{\mathrm{d}\alpha} = \mathrm{tr}\left(\mathrm{adj}(A)\frac{\mathrm{d}A}{\mathrm{d}\alpha}\right).$$

where adj(A) denotes the adjugate of A. In particular, if A is invertible, we have

$$\frac{\mathrm{d}\det(A)}{\mathrm{d}\alpha} = \det(A)\,\mathrm{tr}\left(A^{-1}\frac{\mathrm{d}A}{\mathrm{d}\alpha}\right).$$

Expressed in terms of the entries of A, these are

$$\frac{\partial\det(A)}{\partial A_{ij}} = \mathrm{adj}(A)_{ji} = \det(A)(A^{-1})_{ji}.$$

Yet another equivalent formulation is

$$\det(A + \epsilon X) - \det(A) = \mathrm{tr}(\mathrm{adj}(A)X)\epsilon + O(\epsilon^2)$$
$$= \det(A)\,\mathrm{tr}(A^{-1}X)\epsilon + O(\epsilon^2)$$

using big O notation. The special case where $A = I$, the identity matrix, yields

$$\det(I + \epsilon X) = 1 + \mathrm{tr}(X)\epsilon + O(\epsilon^2).$$

This identity is used in describing the tangent space of certain matrix Lie groups.

If the matrix A is written as $A = \begin{bmatrix} a & b & c \end{bmatrix}$ where a, b, c are vectors, then the gradient over one of the three vectors may be written as the cross product of the other two:

$$\nabla_{\mathbf{a}} \det(A) = \mathbf{b} \times \mathbf{c}$$
$$\nabla_{\mathbf{b}} \det(A) = \mathbf{c} \times \mathbf{a}$$
$$\nabla_{\mathbf{c}} \det(A) = \mathbf{a} \times \mathbf{b}.$$

Abstract Algebraic Aspects

Determinant of An Endomorphism

The above identities concerning the determinant of a products and inverses of matrices imply that similar matrices have the same determinant: two matrices A and B are similar, if there exists an invertible matrix X such that $A = X^{-1}BX$. Indeed, repeatedly applying the above identities yields

$$\det(A) = \det(X)^{-1} \det(BX) = \det(X)^{-1} \det(B) \det(X) =$$
$$\det(B) \det(X)^{-1} \det(X) = \det(B).$$

The determinant is therefore also called a similarity invariant. The determinant of a linear transformation

$$T : V \to V$$

for some finite dimensional vector space V is defined to be the determinant of the matrix describing it, with respect to an arbitrary choice of basis in V. By the similarity invariance, this determinant is independent of the choice of the basis for V and therefore only depends on the endomorphism T.

Exterior Algebra

The determinant can also be characterised as the unique function

$$D : M_n(K) \to K$$

from the set of all n-by-n matrices with entries in a field K to this field satisfying the following three properties: first, D is an n-linear function: considering all but one column of A fixed, the determinant is linear in the remaining column, that is

$$D(v_1, \ldots, v_{i-1}, av_i + bw, v_{i+1}, \ldots, v_n) = aD(v_1, \ldots, v_{i-1}, v_i, v_{i+1}, \ldots, v_n)$$
$$+ bD(v_1, \ldots, v_{i-1}, w, v_{i+1}, \ldots, v_n)$$

for any column vectors v_1, ..., v_n, and w and any scalars (elements of K) a and b. Second, D is an alternating function: for any matrix A with two identical columns $D(A) = 0$. Finally, $D(I_n) = 1$. Here I_n is the identity matrix.

This fact also implies that every other n-linear alternating function F: $M_n(K) \to K$ satisfies

$$F(M) = F(I)D(M).$$

The last part in fact follows from the preceding statement: one easily sees that if F is nonzero it satisfies $F(I) \neq 0$, and function that associates $F(M)/F(I)$ to M satisfies all conditions of the theorem. The importance of stating this part is mainly that it remains valid if K is any commutative ring rather than a field, in which case the given argument does not apply.

The determinant of a linear transformation $A : V \to V$ of an n-dimensional vector space V can be formulated in a coordinate-free manner by considering the n-th exterior power $\wedge^n V$ of V. A induces a linear map

$$\Lambda^n A : \Lambda^n V \to \Lambda^n V$$

$$v_1 \wedge v_2 \wedge \ldots \wedge v_n \mapsto Av_1 \wedge Av_2 \wedge \ldots \wedge Av_n.$$

As $\wedge^n V$ is one-dimensional, the map $\wedge^n A$ is given by multiplying with some scalar. This scalar coincides with the determinant of A, that is to say

$$(\Lambda^n A)(v_1 \wedge \ldots \wedge v_n) = \det(A) \cdot v_1 \wedge \ldots \wedge v_n.$$

This definition agrees with the more concrete coordinate-dependent definition. This follows from the characterisation of the determinant given above. For example, switching two columns changes the parity of the determinant; likewise, permuting the vectors in the exterior product $v_1 \neq v_2 \neq \ldots \neq v_n$ to $v_2 \neq v_1 \neq v_3 \neq \ldots \neq v_n$, say, also alters the parity.

For this reason, the highest non-zero exterior power $\wedge^n(V)$ is sometimes also called the determinant of V and similarly for more involved objects such as vector bundles or chain complexes of vector spaces. Minors of a matrix can also be cast in this setting, by considering lower alternating forms $\wedge^k V$ with $k < n$.

Square matrices over commutative rings and abstract properties The determinant of a matrix can be defined, for example using the Leibniz formula, for matrices with entries in any commutative ring.

Briefly, a ring is a structure where addition, subtraction, and multiplication are defined. The commutativity requirement means that the product does not depend on the order of the two factors, i.e.,

$$r \cdot s = s \cdot r$$

is supposed to hold for all elements r and s of the ring. For example, the integers form a commutative ring.

Many of the above statements and notions carry over mutatis mutandis to determinants of these more general matrices: the determinant is multiplicative in this more general situation, and Cramer's rule also holds.

A square matrix over a commutative ring R is invertible if and only if its determinant is a unit in R, that is, an element having a (multiplicative) inverse. (If R is a field, this latter condition is equivalent to the determinant being nonzero, thus giving back the above characterisation.) For example, a matrix A with entries in Z, the integers, is invertible (in the sense that the inverse matrix has again integer entries) if the determinant is +1 or −1. Such a matrix is called unimodular.

The determinant defines a mapping

$$\mathrm{GL}_n(R) \rightarrow R^{\times},$$

between the group of invertible $n \times n$ matrices with entries in R and the multiplicative group of units in R. Since it respects the multiplication in both groups, this map is a group homomorphism. Secondly, given a ring homomorphism $f: R \rightarrow S$, there is a map $\mathrm{GL}_n(R) \rightarrow \mathrm{GL}_n(S)$ given by replacing all entries in R by their images under f. The determinant respects these maps, i.e., given a matrix $A = (a_{i,j})$ with entries in R, the identity

$$f(\det((a_{i,j}))) = \det((f(a_{i,j})))$$

holds. For example, the determinant of the complex conjugate of a complex matrix (which is also the determinant of its conjugate transpose) is the complex conjugate of its determinant, and for integer matrices: the reduction modulo m of the determinant of such a matrix is equal to the determinant of the matrix reduced modulo m (the latter determinant being computed using modular arithmetic). In the more high-brow parlance of category theory, the determinant is a natural transformation between the two functors GL_n and $(Å–)^{\times}$. Adding yet another layer of abstraction, this is captured by saying that the determinant is a morphism of algebraic groups, from the general linear group to the multiplicative group,

$$\det : \mathrm{GL}_n \to \mathbb{G}_m.$$

Generalisations and Related Notions

Infinite Matrices

For matrices with an infinite number of rows and columns, the above definitions of the determinant do not carry over directly. For example, in Leibniz' formula, an infinite sum (all of whose terms are infinite products) would have to be calculated. Functional analysis provides different extensions of the determinant for such infinite-dimensional situations, which however only work for particular kinds of operators.

The Fredholm determinant defines the determinant for operators known as trace class operators by an appropriate generalisation of the formula

$$\det(I + A) = \exp(\mathrm{tr}(\log(I + A))).$$

Another infinite-dimensional notion of determinant is the functional determinant.

Notions of Determinant Over Non-Commutative Rings

For square matrices with entries in a non-commutative ring, there are various difficulties in defining determinants in a manner analogous to that for commutative rings.

A meaning can be given to the Leibniz formula provided the order for the product is specified, and similarly for other ways to define the determinant, but non-commutativity then leads to the loss of many fundamental properties of the determinant, for instance the multiplicative property or the fact that the determinant is unchanged under transposition of the matrix.

Over non-commutative rings, there is no reasonable notion of a multilinear form (if a bilinear form exists with a regular element of R as value on some pair of arguments, it can be used to show that all elements of R commute). Nevertheless various notions of non-commutative determinant have been formulated, which preserve some of the properties of determinants, notably quasideterminants and the Dieudonné determinant.

Further Variants

Determinants of matrices in superrings (that is, Z/2-graded rings) are known as Berezinians or superdeterminants. The permanent of a matrix is defined as the determinant, except that the factors sgn(σ) occurring in Leibniz' rule are omitted. The immanant generalises both by introducing a character of the symmetric group S_n in Leibniz' rule.

Calculation

Determinants are mainly used as a theoretical tool. They are rarely calculated explicitly in numerical linear algebra, where for applications like checking invertibility and finding eigenvalues the determinant has largely been supplanted by other techniques.

Nonetheless, explicitly calculating determinants is required in some situations, and different methods are available to do so. Naive methods of implementing an algorithm to compute the determinant include using Leibniz' formula or Laplace's formula.

Both these approaches are extremely inefficient for large matrices, though, since the number of required operations grows very quickly: it is of order $n!$ (n factorial) for an $n \times n$ matrix M. For example, Leibniz' formula requires to calculate $n!$ products. Therefore, more involved techniques have been developed for calculating determinants.

Decomposition Methods

Given a matrix A, some methods compute its determinant by writing A as a product of matrices whose determinants can be more easily computed. Such techniques are referred to as decomposition methods.

Examples include the LU decomposition, the QR decomposition or the Cholesky decomposition (for positive definite matrices). These methods are of order $O(n^3)$, which is a significant improvement over $O(n!)$

The LU decomposition expresses A in terms of a lower triangular matrix L, an upper triangular matrix U and a permutation matrix P:

$$A = PLU.$$

The determinants of L and U can be quickly calculated, since they are the products of the respective diagonal entries. The determinant of P is just the sign ε of the corresponding permutation (which is +1 for an even number of permutations and is -1 for an uneven number of permutations). The determinant of A is then

$$\det(A) = \varepsilon \det(L) \cdot \det(U),$$

Moreover, the decomposition can be chosen such that L is a unitriangular matrix and therefore has determinant 1, in which case the formula further simplifies to

$$\det(A) = \varepsilon \det(U).$$

Chapter 13

Algebraic Topology

Algebraic topology is a branch of mathematics which uses tools from abstract algebra to study topological spaces. The basic goal is to find algebraic invariants that classify topological spaces up to homeomorphism, though usually most classify up to homotopy equivalence.

Although algebraic topology primarily uses algebra to study topological problems, using topology to solve algebraic problems is sometimes also possible. Algebraic topology, for example, allows for a convenient proof that any subgroup of a free group is again a free group.

The Method of Algebraic Invariants

An older name for the subject was combinatorial topology, implying an emphasis on how a space X was constructed from simpler ones (the modern standard tool for such construction is the CW-complex). The basic method now applied in algebraic topology is to investigate spaces via algebraic invariants by mapping them, for example, to groups which have a great deal of manageable structure in a way that respects the relation of homeomorphism (or more general homotopy) of spaces. This allows one to recast statements about topological spaces into statements about groups, which are often easier to prove.

Two major ways in which this can be done are through fundamental groups, or more generally homotopy theory, and through homology and cohomology groups. The fundamental groups give us basic information about the structure of a topological space, but they are often nonabelian and can be difficult to work with.

The fundamental group of a (finite) simplicial complex does have a finite presentation. Homology and cohomology groups, on the other hand, are abelian and in many important cases finitely generated. Finitely generated abelian groups are completely classified and are particularly easy to work with.

Setting in Category Theory

In general, all constructions of algebraic topology are functorial; the notions of category, functor and natural transformation originated here.

Fundamental groups and homology and cohomology groups are not only *invariants* of the underlying topological space, in the sense that two topological spaces which are homeomorphic have the same associated groups, but their associated morphisms also correspond — a continuous mapping of spaces induces a group homomorphism on the associated groups, and these homomorphisms can be used to show non-existence (or, much more deeply, existence) of mappings.

Results on Homology

Several useful results follow immediately from working with finitely generated abelian groups. The free rank of the n-th homology group of a simplicial complex is equal to the n-th Betti number, so one can use the homology groups of a simplicial complex to calculate its Euler-Poincaré characteristic.

As another example, the top-dimensional integral homology group of a closed manifold detects orientability: this group is isomorphic to either the integers or 0, according as the manifold is orientable or not. Thus, a great deal of topological information is encoded in the homology of a given topological space.

Beyond simplicial homology, which is defined only for simplicial complexes, one can use the differential structure of smooth manifolds via de Rham cohomology, or Èech or sheaf cohomology to investigate the solvability of differential equations defined on the manifold in question. De Rham showed that all of these approaches were interrelated and that, for a closed, oriented manifold, the Betti numbers derived through simplicial homology were the same Betti numbers as those derived through de Rham cohomology. This was extended in the 1950s, when Eilenberg and Steenrod generalized this approach. They defined homology and cohomology as functors equipped with natural transformations subject to certain axioms (e.g., a weak equivalence of spaces passes to an isomorphism of homology groups), verified that all existing (co)homology theories satisfied these axioms, and then proved that such an axiomatization uniquely characterized the theory.

A new approach uses a functor from filtered spaces to crossed complexes defined directly and homotopically using relative homotopy groups; a higher homotopy van Kampen theorem proved for this functor enables basic results in algebraic topology, especially on the border between homology and homotopy, to be obtained without using singular homology or simplicial approximation.

This approach is also called nonabelian algebraic topology, and generalises to higher dimensions ideas coming from the fundamental group.

Applications of Algebraic Topology

Classic applications of algebraic topology include:

- The Brouwer fixed point theorem: every continuous map from the unit n-disk to itself has a fixed point.
- The n-sphere admits a nowhere-vanishing continuous unit vector field if and only if n is odd. (For $n = 2$, this is sometimes called the "hairy ball theorem".)
- The Borsuk–Ulam theorem: any continuous map from the n-sphere to Euclidean n-space identifies at least one pair of antipodal points.
- Any subgroup of a free group is free. This result is quite interesting, because the statement is purely algebraic yet the simplest proof is topological.

  Namely, any free group G may be realized as the fundamental group of a graph X. The main theorem on covering spaces tells us that every subgroup H of G is the fundamental group of some covering space Y of X; but every such Y is again a graph.

  Therefore its fundamental group H is free. On the other hand this type of application is also handled more simply by the use of covering morphisms of groupoids, and that technique has yielded subgroup theorems not yet proved by methods of algebraic topology.

- Topological combinatorics

Algebraic Topology (Object)

In mathematics, the algebraic topology on the set of group representations from G to a topological group H is the topology of pointwise convergence, i.e. p_i converges to p if the limit of $p_i(g) = p(g)$ for every g in G.

This terminology is often used in the case of the algebraic topology on the set of discrete, faithful representations of a Kleinian group into PSL(2,C). Another topology, the geometric topology (also called the Chabauty topology), can be put on the set of images of the representations, and its closure can include extra Kleinian groups that are not images of points in the closure in the algebraic topology. This fundamental distinction is behind the phenomenon of hyperbolic Dehn surgery and plays an important role in the general theory of hyperbolic 3-manifolds.

Aspherical Space

In topology, a branch of mathematics, an aspherical space is a topological space with all higher homotopy groups equal to 0. If one works with CW complexes, one can reformulate this condition: an aspherical CW complex is a CW complex whose universal cover is contractible.

Indeed, contractibility of a universal cover is the same, by Whitehead's theorem, as asphericality of it. And it is an application of the exact sequence of a fibration that higher homotopy groups of a space and its universal cover are same. (By the same argument, if E is a path-connected space and p: $E \to B$ is any covering map, then E is aspherical if and only if B is aspherical.)

Aspherical spaces are, directly from the definitions, Eilenberg-MacLane spaces. Also directly from the definitions, aspherical spaces are classifying spaces of their fundamental groups.

Examples

- Using the second of above definitions we easily see that all orientable compact surfaces of genus greater than

0 are aspherical (as they have either the Euclidean plane or the hyperbolic plane as a universal cover).

- It follows that all non-orientable surfaces, except the real projective plane, are aspherical as well, as they can be covered by an orientable surface genus 1 or higher.
- Similarly, a product of any number of circles is aspherical.
- Any hyperbolic 3-manifold is, by definition, covered by the hyperbolic 3-space H^3, hence aspherical.
- Let $X = G/K$ be a Riemannian symmetric space of negative type, and Γ be a lattice in G that acts freely on X. Then the locally symmetric space $\Gamma \backslash G / K$ is aspherical.
- The Bruhat-Tits building of a simple algebraic group over a field with a discrete valuation is aspherical.
- The complement of a knot in S^3 is aspherical, by the sphere theorem
- Metric spaces with nonpositive curvature in the sense of Aleksandrov (locally CAT(0) spaces) are aspherical. In the case of Riemannian manifolds, this follows from the Cartan–Hadamard theorem, which has been generalized to geodesic metric spaces by Gromov and Ballmann. This class of aspherical spaces subsumes all the previously given examples.
- Any nilmanifold is aspherical.

Symplectically Aspherical Manifolds

If one deals with symplectic manifolds, the meaning of "aspherical" is a little bit different. Specifically, we say that a symplectic manifold (M,Ω) is symplectically aspherical if and only if

$$\int_{S^2} f^*\omega = \langle c_1(TM), f_*[S^2]\rangle = 0$$

for every continuous mapping

$$f : S^2 \to M,$$

where $c_1(TM)$ denotes the first Chern class of an almost complex structure which is compatible with Ω.

By Stokes' theorem, we see that symplectic manifolds which are aspherical are also symplectically aspherical manifolds. However, there do exist symplectically aspherical manifolds which are not aspherical spaces.

Some references drop the requirement on c_1 in their definition of "symplectically aspherical." However, it is more common for symplectic manifolds satisfying only this weaker condition to be called "weakly exact."

Shape Theory (Mathematics)

Shape theory is a branch of the mathematical field of topology. Homotopy theory is not appropriate for spaces with bad local properties, hence the need for replacement of homotopy theory by a more sophisticated approach.

Thus the overall goal of shape theory is to modify the methods and results from homotopy theory for spaces with good local properties, like CW complexes, to more general compact metric spaces or compact Hausdorff spaces with possibly bad local properties.

Shape theory was founded by the Polish mathematician Karol Borsuk in 1968. Borsuk lived and worked in Warsaw, hence the name of one of the fundamental examples of the area, the Warsaw circle. This is a compact subset of the plane produced by "closing up" a topologist's sine curve with an arc. It has homotopy groups isomorphic to those of a point, but is not homotopy equivalent to it; Whitehead's theorem does not apply because the Warsaw circle is not a CW complex.

Borsuk's original shape theory has been replaced by a more systematic approach by inverse systems, pioneered by Sibe Mardešiæ, and independently, by Timothy Porter. In abstract terms, one starts with a dense subcategory of good objects, and approximates general objects by inverse systems of good objects in the best way in the sense of certain universality property. Thus the object is replaced by pro-object in dense category in appropriate way.

For some purposes, like dynamical systems, more sophisticated invariants were developed under the name strong shape. Generalizations to noncommutative geometry, e.g. the shape theory for operator algebras have been found.

Cone (Topology)

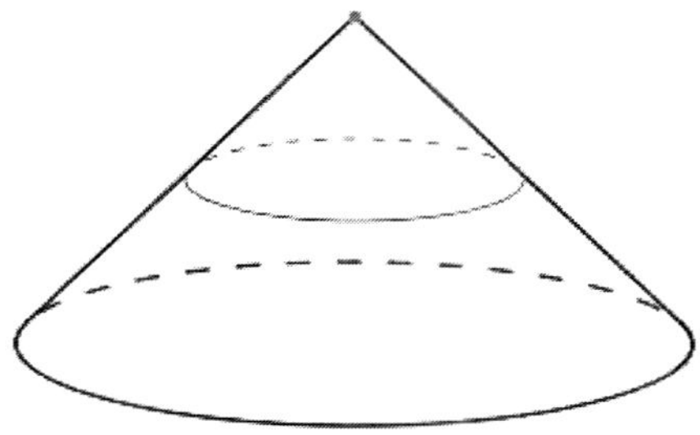

Cone of a circle. The original space is in blue, and the collapsed end point is in green.

In topology, especially algebraic topology, the cone CX of a topological space X is the quotient space:

$$CX = (X \times I) / (X \times \{0\})$$

of the product of X with the unit interval $I = [0, 1]$. Intuitively we make X into a cylinder and collapse one end of the cylinder to a point.

If X sits inside Euclidean space, the cone on X is homeomorphic to the union of lines from X to another point. That is, the topological cone agrees with the geometric cone when defined. However, the topological cone construction is more general.

Examples

- The cone over a point p of the real line is the interval $\{p\} \times [0,1]$.
- The cone over two points $\{0,1\}$ is a "V" shape with endpoints at $\{0\}$ and $\{1\}$.
- The cone over an interval I of the real line is a filled-in triangle, otherwise known as a 2-simplex.
- The cone over a polygon P is a pyramid with base P.
- The cone over a disk is the solid cone of classical geometry (hence the concept's name).
- The cone over a circle is the curved surface of the solid cone:

$$\{(x, y, z) \in \mathbb{R}^3 \mid x^2 + y^2 = z^2 \text{ and } 0 \le z \le 1\}.$$

This in turn is homeomorphic to the closed disc.

- In general, the cone over an n-sphere is homeomorphic to the closed $(n+1)$-ball.
- The cone over an n-simplex is an $(n+1)$-simplex.

Properties

All cones are path-connected since every point can be connected to the vertex point. Furthermore, every cone is contractible to the vertex point by the homotopy

$$h_t(x,s) = (x,\ (1-t)s).$$

The cone is used in algebraic topology precisely because it embeds a space as a subspace of a contractible space.

When X is compact and Hausdorff (essentially, when X can be embedded in Euclidean space), then the cone CX can be visualized as the collection of lines joining every point of X to a single point. However, this picture fails when X is not compact or not Hausdorff, as generally the quotient topology on CX will be finer than the set of lines joining X to a point.

Reduced Cone

If (X, x_0) is a pointed space, there is a related construction, the reduced cone, given by

$$X\times[0,1]/(X\times\{0\})\cup(\{x_0\}\times[0,1])$$

With this definition, the natural inclusion $x\mapsto(x,1)$ becomes a based map, where we take $(x_0,0)$ to be the basepoint of the reduced cone.

Cone Functor

The map $X\mapsto CX$ induces a functor $C:\text{Top}\to\text{Top}$ on the category of topological spaces Top.

Geometry and Topology

In mathematics, geometry and topology is an umbrella term for geometry and topology, as the line between these two is often blurred, most visibly in local to global theorems in Riemannian geometry, and results like the Gauss–Bonnet theorem and Chern–Weil theory.

Sharp distinctions between geometry and topology can be drawn, however, as discussed below. It is also the title of a journal *Geometry & Topology* that covers these topics.

Scope

It is distinct from "geometric topology", which more narrowly involves applications of topology to geometry.

It includes:

- Differential geometry and topology
- Geometric topology (including low-dimensional topology and surgery theory)

It does not include such parts of algebraic topology as homotopy theory, but some areas of geometry and topology (such as surgery theory, particularly algebraic surgery theory) are heavily algebraic.

Distinction Between Geometry and Topology

Pithily, geometry has *local* structure (or infinitesimal), while topology only has *global* structure. Alternatively, geometry has *continuous* moduli, while topology has *discrete* moduli. By examples, an example of geometry is Riemannian geometry, while an example of topology is homotopy theory. The study of metric spaces is geometry, the study of topological spaces is topology.

The terms are not used completely consistently: symplectic manifolds are a boundary case, and coarse geometry is global, not local.

Local Versus Global Structure

Differentiable manifolds (of a given dimension) are all locally diffeomorphic (by definition), so there are no local invariants to a differentiable structure (beyond dimension). So differentiable structures on a manifold is an example of topology. By contrast, the curvature of a Riemannian manifold is a local (indeed, infinitesimal) invariant (and is the only local invariant under isometry).

Moduli

If a structure has a discrete moduli (if it has no deformations, or if a deformation of a structure is isomorphic to the original structure), the structure is said to be rigid, and its study (if it is a geometric or topological structure) is topology. If have non-trivial deformations, the structure is said to be flexible, and its study is geometry.

The space of homotopy classes of maps is discrete, so studying maps up to homotopy is topology. Similarly, differentiable structures on a manifold is usually a discrete space, and hence an example of topology, but exotic R^4s have continuous moduli of differentiable structures. Algebraic varieties have continuous moduli spaces, hence their study is algebraic geometry. Note that these are finite-dimensional moduli spaces. The space of Riemannian metrics on a given differentiable manifold is an infinite-dimensional space.

Symplectic Manifolds

Symplectic manifolds are a boundary case, and parts of their study are called symplectic topology and symplectic geometry. By Darboux's theorem, a symplectic manifold has no local structure, which suggests that their study be called topology. By contrast, the space of symplectic structures on a manifold form a continuous moduli, which suggests that their study be called geometry. However, up to isotopy, the space of symplectic structures is discrete (any family of symplectic structures are isotopic).

Topological Degree Theory

In mathematics, topological degree theory is a generalization of the winding number of a curve in the complex plane. It can be used to estimate the number of solutions of an equation, and is closely connected to fixed-point theory. When one solution of an equation is easily found, degree theory can often be used to prove existence of a second, nontrivial, solution. There are different types of degree for different types of maps: e.g. for maps between Banach spaces there is the Brouwer degree in R^n, the Leray-Schauder degree for compact mappings in normed spaces, the coincidence degree and various other types. There is also a degree for continuous maps between manifolds.

Topological degree theory has applications in complementarity problems, differential equations, differential inclusions and dynamical systems.

Chapter 14

Elementary Function

In mathematics, an elementary function is a function of one variable built from a finite number of exponentials, logarithms, constants, and nth roots through composition and combinations using the four elementary operations (+ – × χ). By allowing these functions (and constants) to be complex numbers, trigonometric functions and their inverses become included in the elementary functions.

The roots of equations are the functions implicitly defined as solving a polynomial equation with constant coefficients. For polynomials of degree four and smaller there are explicit formulae for the roots (the formulae are elementary functions). Elementary functions were introduced by Joseph Liouville in a series of papers from 1833 to 1841. An algebraic treatment of elementary functions was started by Joseph Fels Ritt in the 1930s.

Examples:

Examples of elementary functions include:

$$\frac{e^{\tan(x)}}{1+x^2}\sin\left(\sqrt{1+\ln^2 x}\right)$$

and

$$-i\ln(x+i\sqrt{1-x^2})\cdot$$

This last function is equal to the inverse cosine trigonometric function $\arccos(x)$ in the entire complex domain. Hence, $\arccos(x)$ is an elementary function. An example of a function that is *not* elementary is the error function

$$\mathrm{erf}(x) = \frac{2}{\sqrt{\pi}} \int_0^x e^{-t^2}\, dt,$$

a fact that cannot be seen directly from the definition of elementary function but can be proven using the Risch algorithm.

Differential Algebra

The mathematical definition of an elementary function, or a function in elementary form, is considered in the context of differential algebra. A differential algebra is an algebra with the extra operation of derivation (algebraic version of differentiation).

Using the derivation operation new equations can be written and their solutions used in extensions of the algebra. By starting with the field of rational functions, two special types of transcendental extensions (the logarithm and the exponential) can be added to the field building a tower containing elementary functions.

A differential field F is a field F_0 (rational functions over the rationals Q for example) together with a derivation map $u \to \partial u$. (Here ∂u is a new function. Sometimes the notation $u2$ is used.) The derivation captures the properties of differentiation, so that for any two elements of the base field, the derivation is linear

$$\partial(u+v) = \partial u + \partial v$$

and satisfies the Leibniz product rule

$$\partial(u \cdot v) = \partial u \cdot v + u \cdot \partial v.$$

An element h is a constant if $\partial h = 0$. If the base field is over the rationals, care must be taken when extending the field to add the needed transcendental constants.

A function u of a differential extension $F[u]$ of a differential field F is an elementary function over F if the function u

- is algebraic over F, or
- is an exponential, that is, $\partial u = u\ \partial a$ for $a \in F$, or
- is a logarithm, that is, $\partial u = \partial a / a$ for $a \in F$.

(this is Liouville's theorem).

Exponential Function

In mathematics, the exponential function is the function e^x, where e is the number (approximately 2.718281828) such that the function e^x is its own derivative. The exponential function is used to model a relationship in which a constant change in the independent variable gives the same proportional change (i.e. percentage increase or decrease) in the dependent variable. The function is often written as $\exp(x)$, especially when it is impractical to write the independent variable as a superscript.

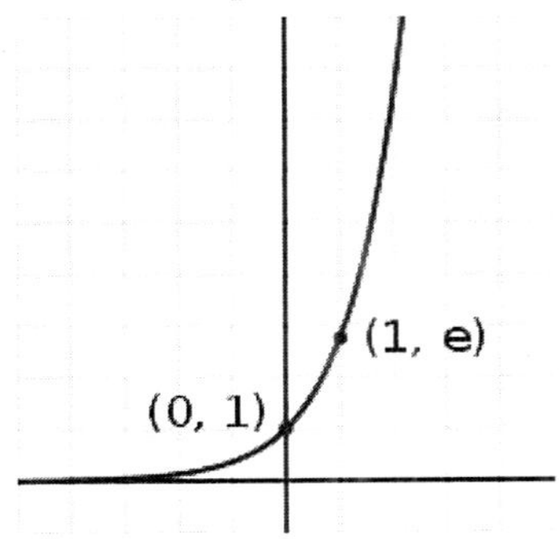

Figure: *The natural exponential function* $y = e^x$

Exponential Function	
Representation	e^x
Inverse	$\ln x$
Derivative	e^x
Indefinite Integral	$e^x + C$

The graph of $y = e^x$ is upward-sloping, and increases faster as x increases. The graph always lies above the x-axis but can get arbitrarily close to it for negative x; thus, the x-axis is a horizontal asymptote. The slope of the tangent to the graph at each point is equal to its y coordinate at that point. The inverse function is the natural logarithm $\ln(x)$; because of this, some old texts refer to the exponential function as the antilogarithm. Sometimes the term exponential function is used more generally for functions of the form cb^x, where the base b is any positive real number, not necessarily e.

In general, the variable x can be any real or complex number, or even an entirely different kind of mathematical object.

Overview

The exponential function arises whenever a quantity grows or decays at a rate proportional to its current value. One such situation is continuously compounded interest, and in fact it was this that led Jacob Bernoulli in 1683 to the number

$$\lim_{n\to\infty}\left(1+\frac{1}{n}\right)^n,$$

now known as e. Later, in 1697, Johann Bernoulli studied the calculus of the exponential function.

If a principal amount of 1 earns interest at an annual rate of x compounded monthly, then the interest earned each month is $x/12$ times the current value, so each month the total value is multiplied by $(1+x/12)$, and the value at the end of the year is $(1+x/12)^{12}$. If instead interest is compounded daily, this becomes $(1+x/365)^{365}$.

Letting the number of time intervals per year grow without bound leads to the limit definition of the exponential function,

$$\exp(x)=\lim_{n\to\infty}\left(1+\frac{x}{n}\right)^n,$$

first given by Euler. This is one of a number of characterizations of the exponential function; others involve series or differential equations.

From any of these definitions it can be shown that the exponential function obeys the basic exponentiation identity,

$$\exp(x+y)=\exp(x)\cdot\exp(y),$$

which is why it can be written as e^x.

The derivative (rate of change) of the exponential function is the exponential function itself. More generally, a function with a rate of change *proportional* to the function itself (rather than equal to it) is expressible in terms of the exponential function. This function property leads to exponential growth and exponential decay.

The exponential function extends to an entire function on the complex plane. Euler's formula relates its values at purely imaginary arguments to trigonometric functions. The exponential function also has analogues for which the argument is a matrix, or even an element of a Banach algebra or a Lie algebra.

Formal Definition

The exponential function e^x can be characterized in a variety of equivalent ways. In particular it may be defined by the following power series:

$$e^x = \sum_{n=0}^{\infty} \frac{x^n}{n!} = 1 + x + \frac{x^2}{2!} + \frac{x^3}{3!} + \frac{x^4}{4!} + \cdots.$$

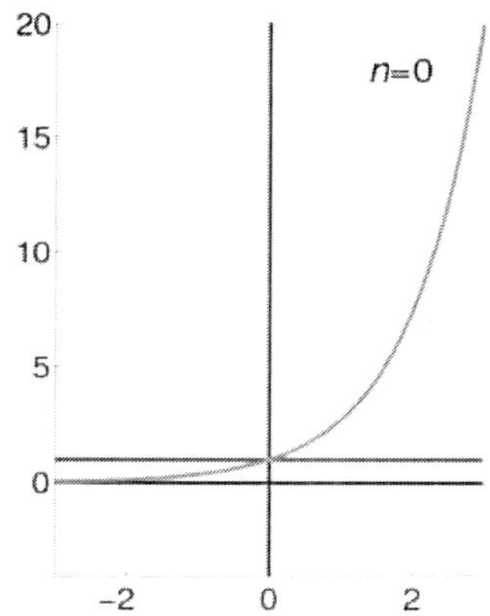

Figure: *The exponential function (in blue), and the sum of the first* n + 1 *terms of the power series on the left (in red).*

Using an alternate definition for the exponential function leads to the same result when expanded as a Taylor series.

Less commonly, e^x is defined as the solution y to the equation

$$x = \int_1^y \frac{dt}{t}.$$

It is also the following limit:

$$e^x = \lim_{n \to \infty} \left(1 + \frac{x}{n}\right)^n.$$

Derivatives and Differential Equations

The importance of the exponential function in mathematics and the sciences stems mainly from properties of its derivative. In particular,

$$\frac{d}{dx}e^x = e^x.$$

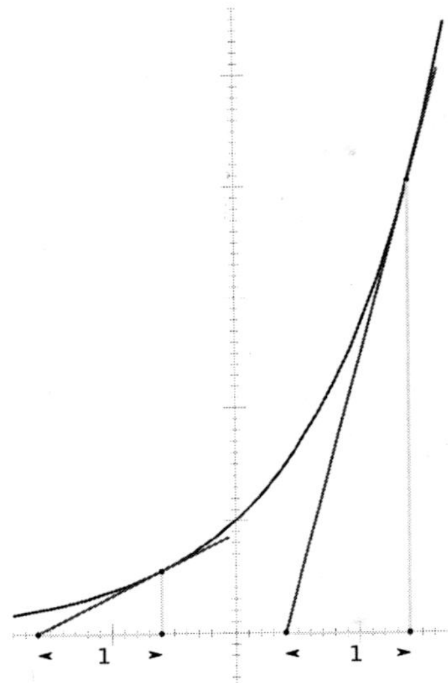

Figure: *The derivative of the exponential function is equal to the value of the function. From any point P on the curve (blue), let a tangent line (red), and a vertical line (green) with height h be drawn, forming a right triangle with a base b on the x-axis. Since the slope of the red tangent line (the derivative) at P is equal to the ratio of the triangle's height to the triangle's base (rise over run), and the derivative is equal to the value of the function, h must be equal to the ratio of h to b. Therefore the base b must always be 1.*

That is, e^x is its own derivative and hence is a simple example of a Pfaffian function. Functions of the form ce^x for constant c are the only functions with that property (by the Picard–Lindelof theorem). Other ways of saying the same thing include:

- The slope of the graph at any point is the height of the function at that point.
- The rate of increase of the function at x is equal to the value of the function at x.

- The function solves the differential equation $y\ 2 = y$.
- exp is a fixed point of derivative as a functional.

If a variable's growth or decay rate is proportional to its size—as is the case in unlimited population growth, continuously compounded interest, or radioactive decay—then the variable can be written as a constant times an exponential function of time. Explicitly for any real constant k, a function f: R→R satisfies $f2\ = kf$ if and only if $f(x) = ce^{kx}$ for some constant c.

Furthermore for any differentiable function $f(x)$, we find, by the chain rule:

$$\frac{d}{dx}e^{f(x)} = f'(x)e^{f(x)}.$$

Continued Fractions for e^x

A continued fraction for e^x can be obtained via an identity of Euler:

$$e^x = 1 + \cfrac{x}{1 - \cfrac{x}{x + 2 - \cfrac{2x}{x + 3 - \cfrac{3x}{x + 4 - \cfrac{4x}{x + 5 - \cfrac{5x}{x + 6 - \ddots}}}}}}$$

The following generalized continued fraction for $e^{2x/y}$ converges more quickly:

$$e^{2x/y} = 1 + \cfrac{2x}{y - x + \cfrac{x^2}{3y + \cfrac{x^2}{5y + \cfrac{x^2}{7y + \cfrac{x^2}{9y + \cfrac{x^2}{11y + \cfrac{x^2}{13y + \ddots}}}}}}}$$

with a special case for $x = y = 1$:

$$e^2 = 7 + \cfrac{2}{5 + \cfrac{1}{7 + \cfrac{1}{9 + \cfrac{1}{11 + \cfrac{1}{13 + \ddots}}}}}$$

Complex Plane

As in the real case, the exponential function can be defined on the complex plane in several equivalent forms. One such definition parallels the power series definition for real numbers, where the real variable is replaced by a complex one:

$$e^z = \sum_{n=0}^{\infty} \frac{z^n}{n!}$$

The exponential function is periodic with imaginary period $2\pi i$ and can be written as

$$e^{a+bi} = e^a(\cos b + i \sin b)$$

where a and b are real values and on the right the real functions must be used if used as a definition. This formula connects the exponential function with the trigonometric functions and to the hyperbolic functions.

When considered as a function defined on the complex plane, the exponential function retains the properties

- $e^{z+w} = e^z e^w$
- $e^0 = 1$
- $e^z \neq 0$
- $\frac{d}{dz} e^z = e^z$
- $(e^z)^n = e^{nz}, n \in \mathbb{Z}$

for all z and w.

The exponential function is an entire function as it is holomorphic over the whole complex plane. It takes on every complex number excepting 0 as value. This is an example of

Picard's little theorem that any non-constant entire function takes on every complex number as value with at most one value excepted. Extending the natural logarithm to complex arguments yields the complex logarithm log z, which is a multi-valued function.

We can then define a more general exponentiation:

$$z^w = e^{w \log z}$$

for all complex numbers z and w.

This is also a multi-valued function, even when z is real. This distinction is problematic, as the multi-valued functions $\log z$ and z^w are easily confused with their single-valued equivalents when substituting a real number for z. The rule about multiplying exponents for the case of positive real numbers must be modified in a multi-valued context:

$$(e^z)^w \neq e^{zw}, \text{ but rather } (e^z)^w = e^{(z+2\pi in)w}$$

multivalued over integers n.

The exponential function maps any line in the complex plane to a logarithmic spiral in the complex plane with the centre at the origin. Two special cases might be noted: when the original line is parallel to the real axis, the resulting spiral never closes in on itself; when the original line is parallel to the imaginary axis, the resulting spiral is a circle of some radius.

Computation of a^b Where both a and b are Complex

Complex exponentiation a^b can be defined by converting a to polar coordinates and using the identity $(e^{\ln(a)})^b = a^b$:

$$a^b = (re^{\theta i})^b = (e^{\ln(r)+\theta i})^b = e^{(\ln(r)+\theta i)b}.$$

However, when b is not an integer, this function is multivalued, because θ is not unique.

Matrices and Banach Algebras

The power series definition of the exponential function makes sense for square matrices (for which the function is called the matrix exponential) and more generally in any Banach algebra B. In this setting, $e^0 = 1$, and e^x is invertible with inverse $\geq^x$ for any x in B. If $xy = yx$, then $e^{x+y} = e^x e^y$, but this identity can fail for noncommuting x and y.

Some alternative definitions lead to the same function. For instance, e^x can be defined as $\lim_{n\to\infty}\left(1+\frac{x}{n}\right)^n$. Or e^x can be defined as $f(1)$, where f: R→B is the solution to the differential equation $f2\ (t) = xf(t)$ with initial condition $f(0) = 1$.

On Lie Algebras

Given a Lie group G and its associated Lie algebra $\mathfrak{g}$, the exponential map is a map $\mathfrak{g} \to G$ satisfying similar properties. In fact, since R is the Lie algebra of the Lie group of all positive real numbers under multiplication, the ordinary exponential function for real arguments is a special case of the Lie algebra situation. Similarly, since the Lie group GL(n,R) of invertible $n \times n$ matrices has as Lie algebra M(n,R), the space of all $n \times n$ matrices, the exponential function for square matrices is a special case of the Lie algebra exponential map.

The identity $\exp(x+y) = \exp(x)\exp(y)$ can fail for Lie algebra elements x and y that do not commute; the Baker–Campbell–Hausdorff formula supplies the necessary correction terms.

Double Exponential Function

The term *double exponential function* can have two meanings:

- a function with two exponential terms, with different exponents
- a function $f(x) = a^{ax}$; this grows even faster than an exponential function; for example, if $a = 10$: $f(-1) = 1.26$, $f(0) = 10$, $f(1) = 10^{10}$, $f(2) = 10^{100}$ = googol, ..., $f(100)$ = googolplex.

Factorials grow faster than exponential functions, but slower than double-exponential functions. Fermat numbers, generated by $F(m) = 2^{2^m} + 1$ and double Mersenne numbers generated by

$$MM(p) = 2^{(2^p-1)} - 1$$

are examples of double exponential functions.

Bibliography

Apostol, Tom M.: *Calculus, Vol. 1: One-Variable Calculus with an Introduction to Linear Algebra,* Wiley, NY, 2007.

Asmussen S.: *Ruin Probabilities*. World Scientific Publishing Co.. Singapore, New York, 2018.

Bartle, Robert G. and Sherbert, Donald R.: *Introduction to Real Analysis,* John Wiley and Sons, New York, 2019.

Boole, George: *A Treatise On The Calculus of Finite Differences*, Macmillan and Company, New York, 2000.

Cartan, E.: *Geometry of Riemannian Spaces,* Math Sci Press, Brookline, MA, USA, 2013.

Courant, Richard: *Differential and Integral Calculus. Vol. II*, Wiley Classics Library, New York: John Wiley & Sons, 2008.

David A. Sprecher: *Elements of Real Analysis,* Dover Publications, Mineola, New York, 2010.

Doolittle, R. F.: *Probability and the Origin of Life,* Scientists Confront Creationists, New York, 2003.

Eisenhart, L. P.: *An Introduction to Differential Geometry with Use of the Tensor Calculus,* Princeton University Press, Princeton, USA, 1990.

Field, Judith V.: *The Geometrical Work of Girard Desargues.* Springer-Verlag, New York, 2007.

Gerald, B. Folland: *Real Analysis: Modern Techniques and Their Applications,* John Wiley & Sons, NY, 2014.

Gerretsen, J. C. H.: *Lectures on Tensor Calculus and Differential Geometry,*Academic Press, New York, USA, 2012.

Henk Tijms: *Understanding Probability*, Cambridge Univ. Press, Cambridge, 2018.

Hildebrand, F.B.: *Introduction to Numerical Analysis,* McGraw-Hill, New York, 2004.

Irene A. Stegun: *Handbook of Mathematical Functions with Formulas, Graphs, and Mathematical Tables*, Dover, New York, 2014.

Jorg Arndt and Christoph Haenel: *Pi—Unleashed,* Springer-Verlag, Berlin, 2015.

Kock, Anders: *Synthetic Differential Geometry,* Cambridge University Press, Cambridge, 2016.

Levi-Civita, T.: *The Absolute Differential Calculus,* Dover Publications, New York, USA, 2007.

Longo, M. and V. Valori: *The Comparison Test — Not Just for Nonnegative Series,* Mathematics Magazine, 2016.

Michael J. Crowe: *A History of Vector Analysis: The Evolution of the Idea of a Vectorial System*, Dover Publications; New York, 2007.

Mlodinow, M.: *Euclid's window (the story of geometry from parallel lines to hyperspace),* Harper & Row, New York, USA, 2004.

Navidi, William: *Statistics for Engineers and Scientists*, McGraw-Hill, New York, 2018.

Nomizu, K. and T. Sasaki: *Affine Differential Geometry,* Cambridge University Press, Cambridge, UK, 1996.

Olav Kallenberg: *Foundations of Modern Probability,* Springer , New York, 2022.

Oprea, John: *Differential Geometry and Applications*, Prentice Hall, Englewood Cliffs, New Jersey, 2023.

Patrick Billingsley: *Probability and Measure*, Toronto, London: John Wiley and Sons, New York, 2009.

Patterson, J. W.: *Thermodynamics and Probability*, Cambridge Univ. Press, Cambridge, 2004,

Richtmeyer, D. and Morton, K.W.: *Difference Methods for Initial Value Problems*, Wiley, New York, 2007.

Robinson, E. A.: *Least Squares Regression Analysis in Terms of Linear Algebra*, Goose Pond Press, Houston, USA, 2011.

Saville, D. J. and G. R. Wood: *Statistical Methods: The Geometric Approach*, Springer-Verlag, New York, USA, 2013.

Schey, H. M.: *Div Grad Curl and all that: An Informal Text on Vector Calculus,* W. W. Norton & Company, UK, 2015.

Translation: Kolmogorov: *Foundations of the Theory of Probability*, Chelsea, New York, 1996.

Wang, Zhenyuan, and, George J. Klir.: *Fuzzy Measure Theory*, Plenum Press, New York, 2011.

Weintraub, S. H.: *Differential Forms: A Complement to Vector Calculus,* Academic Press, San Diego, CA, USA, 2017.

Index

❑❑❑